SCARCITY AND GROWTH CONSIDERING OIL AND ENERGY

An Alternative Neo-Classical View

Front Cover: Trans-Alaska Oil Pipeline, Fairbanks, Alaska.

SCARCITY AND GROWTH CONSIDERING OIL AND ENERGY
An Alternative Neo-Classical View

Douglas B. Reynolds

Symposium Series
Volume 65

The Edwin Mellen Press
Lewiston•Queenston•Lampeter

Library of Congress Cataloging-in-Publication Data

Reynolds, Douglas Bradford.
Scarcity and growth considering oil and energy : an alternative neo-classical view / Douglas B. Reynolds.
p. cm. -- (Symposium series ; v. 65)
Includes bibliographical references and index.
ISBN 0-7734-7234-7
1. Petroleum--Prices--Econometric models. 2. Power resources--Prices--Econometric models. 3. Economic development--Econometric models. 4. Supply and demand--Econometric models. 5. Hubbert, M. King (Marion King), 1903- 6. Hotelling, Harold, 1895- I. Title. II. Symposium series (Edwin Mellen Press) ; v. 65.

HD9560.4 .R49 2002
333.79--dc21

2001044207

This is volume 65 in the continuing series
Symposium Series
Volume 65 ISBN 0-7734-7234-7
SS Series ISBN 0-88946-989-X

A CIP catalog record for this book is available from the British Library.

The Edwin Mellen Press
Box 450
Lewiston, New York
USA 14092-0450

The Edwin Mellen Press
Box 67
Queenston, Ontario
CANADA L0S 1L0

The Edwin Mellen Press, Ltd.
Lampeter, Ceredigion, Wales
UNITED KINGDOM SA48 8LT

Printed in the United States of America

to my loving wife

Candace

Scarcity and Growth Considering Oil and Energy: An Alternative Neo-Classical View

By Douglas B. Reynolds

Table of Contents

Commendatory Preface i

Introduction 1

Chapter 1. 17
The Mineral Economy:
How Prices and Costs Can Falsely Signal Decreasing Scarcity

From Douglas B. Reynolds. (1999) "The Mineral Economy: How Prices and Costs Can Falsely Signal Decreasing Scarcity," *Ecological Economics*, Volume 31, Number 1, pp. 155-166.

Chapter 2. 41
Determining the power of prices to change oil discovery and production using a non-time dependent Hubbert Model

From Douglas B. Reynolds. (2001) "Determining the power of prices to change oil discovery and production using a non-time dependent Hubbert Model," Working paper.

Chapter 3. 57
Oil Exploration Game with Incomplete Information: An Experimental Study

From Douglas B. Reynolds. (2001) "Oil Exploration Game with Incomplete Information: An Experimental Study," *Energy Sources*. Volume 23, Number 3, April, pp. 77-84.

Chapter 4. 69
Modeling OPEC Behavior:
Theories of Risk Aversion for Oil Producer Decisions

From Douglas B. Reynolds. (1999) "Modeling OPEC Behavior: Theories of Risk Aversion for Oil Producer Decisions," *Energy Policy*, Volume 27, pp. 901-912.

Chapter 5. 95
World Oil Supply Forecast and the Reserve-Production Ratio

From Douglas B. Reynolds. (2001) "World Oil Supply Forecast and the Reserve/Production Ratio" *2001: An Energy Odyssey*, Conference Proceedings of the 24th Annual International Conference of the International Association of Energy Economics (IAEE), April 25 –27 Houston, Texas.

Chapter 6. 111
Energy Grades and Economic Growth

From Douglas B. Reynolds. (1994) "Energy Grades and Economic Growth," *Journal of Energy and Development*, Volume 19, Number 2, Spring 1994, pp. 245 – 264.

Chapter 7. 131
Entropy Subsidies

From Douglas B. Reynolds. (1998) "Entropy Subsidies," *Energy Policy*, February, Volume 26, Number 2, pp. 113-118.

Chapter 8. 141
The Energy Utilization Chain: Determining Viable Oil Alternative Technology

From Douglas B. Reynolds. (2000) "Energy Utilization Chain: Determining Viable Oil Alternative Technology," *Energy Sources*, Volume 22, Number 3, April, pp. 215-226.

Chapter 9. 159
The Economics of Alternative Energy Resources

From Douglas B. Reynolds. (2001) "The Economics of Alternative Energy Resources," Working paper.

Chapter 10. 177
Entropy and Diminishing Elasticity of Substitution

From Douglas B. Reynolds. (1999) "Entropy and Diminishing Elasticity of Substitution," *Resources Policy*, March, Volume 25, Number 1, pp. 51 – 58, 25th Anniversary Volume.

Chapter 11. 197
Soviet Economic Decline:
Did an Oil Crisis Cause the Transition in the Soviet Union?

From Douglas B. Reynolds. (2000) "Soviet Economic Decline: Did an oil Crisis Cause the Transition in the Soviet Union," *Journal of Energy and Development*, Volume 24, Number 1, pp. 65-82.

Index. 221

List of Illustrations

Chapter 1. 17

Figure 1.1 Price, production and Hubbert supply trend as a function of cumulative production 35

Figure 1.2 Price and production as a function of time 36

Chapter 2. 41

Figure 2.1 The production Hubbert curve with low oil production 44

Figure 2.2 The discovery and production Hubbert curves 46

Figure 2.3 US lower 48 oil discovery and real oil price as a function of cumulative discovery 50

Table 2.1 US lower 48 oil discovery: econometric analysis of discovery, 1900 to 1999 51

Figure 2.4 US lower 48 oil production and real oil price as a function of cumulative production 52

Table 2.1 US lower 48 oil production: econometric analysis of production, 1900 to 1999 53

Chapter 3. 57

Table 3.1 Oil exploration region and site characteristics 63

Figure 3.1 Discovery, exploration effort and success rates versus cumulative discovery 66

Chapter 4. 69

Figure 4.1 Small firm risk utility curve for exploration of a potential new oil field where the field is large relative to the size of the firm 76

Figure 4.2 Small firm risk utility curve for development of a potential new oil field where the field is large relative to the size of the firm 79

Figure 4.3 Large firm risk utility curve for exploration of a potential new oil field where the field is not large relative to the size of the firm 80

Figure 4.4 Large firm risk utility curve for development of a potential new oil field where the field is not large relative to the size of the firm 81

Figure 4.5 Country monopoly risk utility curve for exploration of a potential new oil field where the field is not large relative to the size of the country GDP 84

Figure 4.6 Country monopoly risk utility curve for development of a potential new oil field where the field is not large relative to the size of the country GDP 87

Chapter 5. 95

Figure 5.1 EIA forecasted supply and demand for world crude oil as a function of cumulative production 102

Figure 5.2 Forecasted OPEC supply and world oil demand as a function of cumulative production 108

Chapter 6. 111

Table 6.1 Energy grades 118

Figure 6.1 Energy sources state grade 119

Figure 6.2 Energy sources ranges of area grades 119

Figure 6.3 Energy sources volume grades 120

Figure 6.4 Energy sources weight grades 120

Figure 6.5 State grade history: 10 million BC to 2000 AD 121

Figure 6.6 Area grade history: 10 million BC to 2000 AD 121

Figure 6.7 Volume grade history: 10 million BC to 2000 AD 122

Figure 6.8 Weight grade history : 10 million BC to 2000 AD 123

Chapter 10. 177

Figure 10.1 Labor-capital isoquant 185

Chapter 11. 197

Figure 11.1 US lower 48 oil production and real oil production as a function of cumulative production 212

Figure 11.2 Soviet and former Soviet oil production 213

Commendatory Preface

The American life style is astonishingly dependant upon cheap energy, and in particular liquid and gaseous petroleum. But most Americans do not appreciate or understand this dependence (other than filling their automobile), because it takes place elsewhere in our economy. For example it takes the energy equivalent of about one gallon of oil per day to feed an American, and of roughly 10 gallons per day to do the equivalent of hewing our wood and hauling our water, that is, to do the basic economic underpinnings of our life. This is the energy equivalent of about 87 barrels of oil per year (of which roughly 40 percent is derived from oil, about 25 percent each from gas and from coal, and the rest from hydro and nuclear power). How long might this energy be available? Can it be increased greatly to meet the expectations of our own and especially the world's growing economies, whose economic growth is tied very close to one for one with increased energy consumption?

It is very hard to predict that for sure, but we can at least make some simple calculations to indicate whether there might be a problem. The proved oil reserves of the United States in 2000 were about 21 billion 42 gallon barrels, which seems like quite a lot. But 280 million people using 87 barrels apiece each year would use 24 billion barrels of oil, so that if our energy requirements were dependant upon only domestic oil (which fortunately is not the case) we would use up our proved reserves in barely more than a year. Clearly there is some serious thinking that needs to be done about our energy reserves and our rate of energy use.

Nevertheless there has been surprisingly little analysis of what the principal determinants of the future availability of such resources might be. Economists, at least those with an historical perspective, tend to consider the work of Hotelling, who predicted that the price of a non-renewable resource (such as oil) would rise gradually over time as the highest grade resources were exhausted until the reserves were depleted. But Barnett and Morse, in a much quoted book in 1963, found little or no increase over time in the price of most resources (once corrected for the general inflation rate). This indicated to Barnett and Morse that non renewable resources such as petroleum were not becoming scarcer to society, probably, they thought, because technical innovations were keeping up with depletion. The implication was that this would continue to be the case, perhaps indefinitely, and most economists stopped worrying about the physical supplies of resources, and have instead focused on other market-related things. At the extreme many economists have, and continue to, dismiss all pronouncements of those concerned about future availability of resources because in their mind all such issues would be resolved by market forces, technology and substitution. At the time of this writing this view is overwhelmingly the prevalent one in the United States. They have, for the time being at least, clearly won the ideological war over the resource constraintists.

In sharp contrast a divergent group of natural scientists and a very few economists have focused instead on the physical requirements for, and possible limits to, economic activity. Chief amongst these was geologist M. King Hubbert, who gained great respectability by predicting in 1955 that U.S. oil production would peak in 1970, which it did. He developed the concept that a natural resource would, over its lifetime, be used in a pattern similar to a normal or bell-shaped curve. To date Hubbert's model continues to be a remarkably good predictor of U.S. oil production. Hubbert also made a similar analysis of global oil production, predicting in 1968 that the global peak in oil production would occur between 1990 and 2000, depending upon the ultimate quantity of extractible oil left in the ground (2 to 3 trillion barrels was his guess). There has been considerable speculation about the degree to which world production of oil would follow a Hubbert curve and if so when it might peak. For example, Campbell (and Campbell and Laherre) predicted that this peak would occur in 2007. Others rushed into print analyses that indicate that the peak would occur at a little to a lot (even 100 years!) later.

If Hubbert was so accurate for the U.S. why does he seem to be missing the mark for global oil production, since total oil production is still increasing into 2001? One answer can be seen by comparing Hubbert's 1968 figure 8.23 and Campbell and Laherrere's empirical figure in their Scientific American "Global production of oil ". Hubbert's curve seems to work pretty well against the subsequent data until about 1980, and then the actual production of oil drops well below Hubbert's prediction. The oil not produced from 1980 to 2000 is roughly that required to extend the peak of production until 2007, assuming, perhaps charitably, that Campbell's own analysis of the peak is correct. What seems to have happened is that, ironically, the economic slowdown of the 1980s, itself caused by the oil price shocks of the 1970s, has decreased oil extraction sufficiently to have saved us from reaching the peak of the Hubbert curve by 2000. The Asian economic "meltdown" of 1998 similarly gave us some breathing room. Thus there is still time for Hubbert to be approximately correct, or perhaps there is so much oil left in the world that his analysis is irrelevant. Meanwhile technical improvements in finding and extracting oil, and international competition, have kept prices low, apparently further negating Hotelling's predictions.

What is curious is that, given the undeniable importance of petroleum to almost everything else in America, why so few analysts have explored the issue in any particular depth recently (The work of Cleveland and Kaufmann is an exception). It seems pretty important to me to judge to what degree these earlier thinker's concepts have been negated or whether their analysis is basically correct, which might lead to some rather nasty impacts on our petroleum-dependant economy. Perhaps it is the rather uncritical acceptance of market economics (after all , some would say, market economics has resolved the energy issues of the 1970s and 1980s) that has prevented a more interesting analysis, perhaps it is the lack of methodological training in biophysical reality (vs. economic theory) in our nation's universities that has caused this, or perhaps it is something else. In any case, we have, unfortunately, a remarkably poor arsenal of new economic or analytic tools to help us understand what kind of oil future we might expect.

The basic fact that all analysts must accept is that the data base needed to calculate essentially any of the parameters necessary to predict the Hubbert (or any other) curve for nearly any other part of the world besides the US are simply not known. For example we still have no idea how much extractable oil is out there because it is not in the interest of, say, Iraq, to tell us, even if they know. There are some disquieting clues. Campbell's 1998 Oil and Gas Journal paper shows the time and magnitude of major discoveries for all major oil producing countries of the world. Basically there have not been *any* large new oil field discoveries in most oil producing countries since about 1970. This is more striking to me than the more frequently quoted prediction of a global oil production peak in 2007, and suggests that we already know roughly how much oil we will find (although our efficiency of extraction may continue to increase, at an unknown energy cost). Another disquieting fact from his analysis is that a substantial portion of our "proved" oil reserves was "found" in OPEC nations during the same year (1988) , indicating to me that they were found by fiat only, under the incentive of OPEC assigning allowable national production in proportion to reserves. Maybe there is much *less* oil out there than even the conservative "proved reserves" suggest. Maybe there is much more.

Given all these uncertainties it is refreshing to see the depth, rigor and new insight that Douglas Reynolds beings to bear on this problem. Reynolds extends our analytic arsenal for understanding these critical questions in a number of very interesting and powerful ways. He makes three principal points, which I agree with completely: 1) oil is special, and represents some kind of qualitative peak in the availability of high quality energy resources for the support of human endeavors 2) energy in general, and oil in particular, is critical to the operation of modern economies in a way much more fundamental than most of us realize 3) increasing the money diverted to the oil companies and increasing the drilling rates are unlikely to make much difference in the availability of oil to the American Economy and 4) despite very large reserves of oil (at least based on the short term perspective) the international availability of oil cannot be taken for granted because as global demand increases, many small producers will run dry, at least with respect to export potential. Thus there will be more and more demands put on the relatively few producers that may have very little interest in increasing the rate at which they extract their known supplies. The net effect of all this is, according to Reynolds, a very uncertain reliability for our future oil requirements, at least with respect to inexpensive oil. Thus, according to Reynolds, Campbell's often quoted "End of cheap oil" in about 2007 might indeed turn out to be the case even if oil is physically abundant. Why? I invite the reader to explore this fascinating book and draw his or her own conclusions.

May 2001

Charles A. Hall
Professor, Systems Ecologist
SUNY College of Environmental Science and Forestry
Syracuse University

Introduction

This book offers an alternative set of theories and evidence about scarcity, contradicting the works Scarcity and Growth (1963), and Scarcity and Growth Reconsidered (1979). The innovation here is to combine basic economic principles of scarcity and cost with physics in a way that has never before been done. A key question in the debate over scarcity is if oil prices have gone down for 100 years, then that should indicate technology has triumphed over scarcity. However, as you will see in the following chapters I show step by step the theory, which explains this phenomenon and why price and costs do not indicate decreasing scarcity. To explain scarcity, I use the concept of "Bonanza" explained by Georgescu-Rogen and the "Mayflower Problem" espoused by Norgaard, as well as M. King Hubbert's famous oil logistics curve. After a through investigation, we will see that the economy can easily be "fooled" into believing that scarcity is declining, when in fact it is increasing rapidly. Eventually the true scarcity of the economic system is revealed in a shock, too late to alter the course of economic growth. Harold Hotelling's (1931) famous Hotelling Principle cannot be used to prepare the economy or the market for sharply higher resource prices because of the lack of information in the early stages of energy resource exploration. In addition uncertainty theory cannot be used effectively to plan for a price shock either because the resource base is never known until the inflection point is reached at which time it is too late to prepare.

Other innovations in the book explain contradictions in many commonly used resource economic models. For example, it is generally thought that any energy resource at all, such as solar or wind power, can be substituted for oil or natural gas. This book explains that such substitutions are extremely difficult to make. The costs of energy substitutions are constrained by the physics of how each energy resource works within an economic system. There is no panacea. No one energy resource can easily and effectively emerge as an adequate substitute to take care of the economy. The inevitable consequence of this situation is inevitable economic decline.

This book also explains a new theory of how and why OPEC works. OPEC is shown not to be a cartel, but rather a set of risk averse oil producers. Their risk averse nature results in lower production. Eventually OPEC's low production and the world's high demand will create another shortage that will emerge quickly and without warning as it did in the 1970's. Interestingly this same situation occurred to the former Soviet Union, where the fall of the Soviet Union was as a result of an oil crisis. I will explain that fall and show what we can learn from it.

Not only is this book academically sound, but it is easily accessible to students who are studying these and other issues. People who have read these articles have been amazed and have radically changed their belief about energy scarcity and how they view the world. Below is a summary and interpretation of the main points of each chapter.

In chapter 1, a new model of resource extraction called the mineral economy model is explained. It is a model in contrast to the usual "corn" and "hardtack" economy models that are used to show how powerful technology is. The difference between the hardtack model and the mineral model is simply that in the mineral model, hardtack is hidden underground making it very inaccessible to resource explorers. The explorers must dig into the ground to find the hardtack. The mineral economy is an alternative theory for explaining how scarcity works.

Ever since Thomas Malthus' and David Ricardo's concepts of scarcity, the economics profession has followed these classic theories without question. Malthusian scarcity is where society extracts and uses resources of a constant quality until one day the resource simply runs out. Ricardian scarcity is a more elegant theory, which states that society extracts and uses resources starting with the highest quality first and proceeding to the lowest quality. Eventually, the quality becomes so low as to be useless in any economic sense. Economists usually use Ricardian scarcity.

Harold Hotelling (1931) states, "The cost of extraction increases as the mine goes deeper," which is the classic Ricardian case. He also states that, "Great wastes arise from the suddenness and unexpectedness of mineral discoveries, leading to wild rushes," and further that, "The problem of exhaustible resources involve the time in another way besides bringing on exhaustion and higher prices, namely as bringing increased information as to the physical extent and condition of the resources." Yet he does not model these ideas on information and discoveries. Instead he assumes them away. That is fine for his paper and theory, but it is not fine if economists use his principle over and over again without accounting for information and discoveries to analyze the economics of resources.

We get a further deepening of the classic Ricardian model into the natural resource economics literature with Barnett and Morse (1963). They write, "For the Ricardian case, we assume … that society possesses both the knowledge and the will to use resources in order of declining economic quality, that is, in an order which results in increasing cost per unit of output." They do speculate somewhat whether this assumption is correct or not when they say, "it is rather much to expect that society will never be surprised by discoveries of hitherto unknown resources." However, they assure us, "None of the … modifications of our initial conditions does material violence to the more general tenets of classical economic theory … a tendency to recurrent upward movement in the cost of output of the extractive sector would *seem* (emphasis added) more probable than a constant or declining trend."

Therefore a lot of the Scarcity and Growth conclusions rest on whether the Ricardian model is always correct. It would be a good idea therefore to test the Ricardian hypothesis, however, Barnett and Morse did not. Their statistical analysis and data proved that if the Ricardian model is correct, then technology has been able to overcome scarcity. On the other hand, their analysis depends very crucially on the Ricardian model. They did a great job with the tools and theories at hand, but if the Ricardian model is not correct, then their conclusions are misleading. Their position is precarious. Barnett (1979) further nails down the same Ricardian theory and the theory that technology can overcome Ricardian scarcity, with a few qualifications, when he says, "The analysis in Scarcity and Growth suggests that the classical theory of increasing economic scarcity of natural resources was invalid ... (because of) ... substitutes, ...increased discoveries,...increased imports,... and a marked increase in ... sociotechnical improvements." In other words, the Ricardian model with technology change is the most relevant theory. Discoveries may be important too. The problem is that Barnett neither models discovery nor explains how imports can also become depleted eventually. Therefore again, the evidence is misleading.

Interestingly, all the theorists who believe the Ricardian model is the only pertinent theory always mention the possibility of "increased discoveries" or "surprises of new discoveries" or "increased information of new discoveries," yet no one seems to pursue these lines of thought. Information is never modeled. How discoveries occur is never modeled. It is as if once a line of reasoning is started in one direction, new directions are difficult to find. However, there are many exceptions. A number of economists and non-economists alike have created alternative theories different from the Ricardian model including Georgescu-Roegen, Norgaard, Uhler, Cleveland, Hubbert and many others. They failed, however, because they either did not create a total coherent theory that others could understand, (i.e. a mathematical model) or they were completely pushed outside of the mainstream. Possibly because the mainstream could not understand them, or maybe because society was not ready to listen, they were not well studied. This book is an attempt to put some harder theory into these alternative ideas. I have benefited as an economist by a strong background in engineering.

Turning back to Hotelling, one of the biggest and most important theories in all of natural resource economics is the Hotelling principle of unfailing increases in resource prices given certain assumptions. However, given that the Hotelling principle has not worked all these years, it has been assumed that technology is more powerful than scarcity and that therefore human technical innovation is more powerful than nature. On this principle rests all of economics. Yet there is an alternative theory to show why technology is not as powerful as we would like to think.

In chapter one, I build an alternative theory of resource extraction called the "mineral" economy where hardtack is hidden underground. The interesting thing is, as the hardtack is found there is more information about where all the hardtack lies making it easier and easier to find. The result is an increasing rate of discovery and extraction. Costs of extraction also decline. Unfortunately, once depletion sets in, the rate of discovery and extraction suddenly and shockingly declines and costs suddenly and shockingly increase. The result of the mineral economy model is that in the early stages of exploration, there is the appearance of a "corn" economy where production increases and costs decrease, but in later stages a sever "hardtack" economy reality emerges.

The books Scarcity and Growth and Scarcity and Growth Reconsidered have strong theories about the power of technology to over come resource shortages. Unfortunately, the proof that is used is based on the early experience of the mineral economy not the depletion part in the later stages. This created a misleading conclusion. Barnett and Morse have used only half of the data to prove their thesis and they used the wrong half, which can easily create such a mistake. The other half of the data is not used like for example US oil production since 1970. Chapter 1 explains an alternative neoclassical theory from theirs using the mineral economy concept and incorporating the work of the famous geologist M.

King Hubbert. Hubbert's work is so important that it would be good to say a little about who he is.

M. King Hubbert (1903 – 1989) has been for a long time and even today a very controversial figure in the natural resource fields. It seems most professionals really hate his ideas or really love them with very little compromise, although lately more and more professionals have gone over to the side of respect for his work. Hubbert was a geologist and worked for oil companies and the US geological survey, a government agency employed to determine the countries mineral reserves. Hubbert's main theory is very simple and elegant but subject to much interpretation. Basically he said that all mineral extraction over time starts low increases to a peak and then declines. While such a proposition may either sound obvious or over simplified, it will be shown to be quite powerful.

Until recently M. King Hubbert's work has gotten very little attention by economists even though his work has been proven correct beyond doubt. The theory presented here uses Hubbert's work and adapts it into the mineral economy model of natural resource extraction. However, here I simplify Hubbert. In other models using Hubbert, Hubbert's work is fit into economics rather than taking economics and fitting it into Hubbert's work. The difference is significant. The result of taking Hubbert's logistics curve or his yield per effort curve and fitting in economic variables is a very complex interaction of different variables and time.

It reminds one of the early theories by astronomers of how the planets move around the sun. They thought the planets followed circles and then followed a circle on top of a circle and then another circle on top of that circle. It was a circles within circles theory. Finally Kepler disproved the circles within circles theory and provided an alternative model of motion using elliptical orbits now known as Kepler's laws. Just so, models using Hubbert's work tend to have many and varied variables added on to the Hubbert logistics curve rather than creating an economic model that produces a typical Hubbert curve. The model here is simplified. It uses two variables, information and depletion, and shows how those variables interact with each other to create a simplified Hubbert curve using cumulative discovery and cumulative discovery squared. Additional forces such as price and technology work with the information and depletion effects to create an over all coherent model.

Hubbert type models are often called life cycle models and are usually very complex. Other mineral extraction models for oil supplies are even more complex such as Hotelling models, flow chart models, yield per effort models, econometrics models, geologic-engineering process models, and decline curve models. While these models certainly have their uses for a given objective, the long-term trend has followed a simplified Hubbert quadratic curve shown here as well as any of them. Therefore, in the theory presented here, I hope to show that a simplified Hubbert curve is more useful and more accurate then any other model. The mineral economy model starts with a simple information and depletion concept that explains Hubbert's curves. It is a simplified life-cycle model.

Chapter 2 uses the theory of chapter 1 to look at US data. The question is how powerful is price for increasing supplies of natural resources in a mineral economy. The answer is very little. Chapter 2 shows clear evidence that the elasticity of supply with respect to price is extremely inelastic, on the order of 0.05 or less, which suggests that if an oil price shock occurs, there will not be much response from the supply side. In other words, the Hubbert curve is an inelastic limit of supply. I call it the Hubbert limit. This puts into question the commonly held belief by economists that prices are very powerful for creating new supplies of non-renewable resources. An economy that depends on a non-renewable energy resource such as oil and that experiences a shortage or a price shock of that resource will be forced to make all its adjustments on the demand side as long as supply is at the Hubbert limit. Supply side economics, i.e. the idea that we can expand supplies with free markets, will not work in a mineral economy. Supplies are limited. Chapter 2 shows this and it also shows that the theory surrounding chapter 1 does work. This suggests that once an oil price shock occurs, there is very little leeway inside the supply market.

Another aspect of the Hubbert curve though is technology and market structural changes. How powerful is technology or how powerful are changes in market structures such as taxes and property rights for changing the Hubbert curve? I show in this chapter that changes in the Hubbert curve are easily modeled with dummy variables where shifts occur. This explains how structural changes in the market and new technology can shift the Hubbert curve. Most of the shifts in the Hubbert curve though are small and they never shift the direction of the supplies upward. The most that technology is able to do is to slow down the decline in oil production and extend the life of the Hubbert curve. But it cannot change the curve. The fundamental shape of the Hubbert curve stays the same for close to 100 years of data.

Chapter 3 does one additional experiment to show that the theory of chapter 1 works. It uses an experimental game with actual players to see if in fact they behave similar to the theory of chapter 1. Players in a simulated oil game are shown to act in the same manner that we see actual oil explorers act in the real world in the US and elsewhere. The results of the game show conclusively that M. King Hubbert's theory, the Hubbert curve, is based on actual economic motives.

This chapter is in contrast to an article and theory posited by Menard and Sherman (1979). They use a statistical model to analyze oil exploration. In their model, they test whether a random exploration plan works better than the actual exploration sequence that occurred in the US. They find that a random exploration plan would discover the super giant East Texas oil field earlier than what actually happened in the US. They conclude that oil exploration was less successful than a random search. The Menard and Sherman analysis however does not take into account the economic motive for exploration and discovery, namely profit and loss. If everyone knew with 100% probability from the beginning of exploration that East Texas actually existed somewhere, then yes, they would all have drilled randomly. But since there was no assurance that East Texas even existed, then random drilling can cause huge losses and that is why random drilling is never done.

Any model that includes business or economics, but does not account for such variables as the profit motive, risk, expectations, interest rates, property rights, markets, and welfare should make it clear why these variables are not used. Risk of loss is an important variable that the Menard and Sherman model does not include. However, it is often the case that Science and Nature analysts touch on market supply and demand and yet do not account for economics. Why would oil explorers not carry out a random search for the East Texas oil field when there was so much profit to be made? Because there were huge potential losses. As the theory of chapter 1 explains, oil explorers use the best information available to maximize benefits and minimize costs. Since they do not know of the existence of the East Texas oil field, they minimize costs by not drilling randomly but by using information that they know about. In no way did oil explorers drill less successfully than a random search when potential losses are considered. This chapter shows an example.

In chapter 4, I present an alternative theory of OPEC. Usually OPEC is considered to be a cartel or some sort of price leading oligopoly. Here I show that another way in which individual OPEC members act is by being very risk averse to oil exploration and development. Any country with a single government owned or controlled oil company for the entire country tends to be politically risk averse and afraid of making mistakes. This causes individual OPEC countries and even non-OPEC countries like Mexico to be very slow at exploration and development causing them to have very low oil production in comparison to free market competitive countries. This means that OPEC is partly a price leading monopoly with Saudi Arabia as the swing producer, but also that OPEC is a set of very risk averse oil producers.

In other words, one reason OPEC countries do not seem to be as competitive as we might normally expect is because each individual country is very reluctant to invest in strong exploration and development programs. They are politically risk averse. Each country's political ruling entity is afraid that if one exploration or development project loses even the slightest bit of money, then their citizens will complain, and their power will be threatened. On the other hand, any gains of exploration and development get dissipated in government programs so citizens don't see any gains. Citizens see loses, not gains. Citizens react very harshly to any losses by their government because they feel it is their own personal loss. However, citizens do not react very happily when the government has gains, because the citizens do not really see those gains and they feel like that is what the government is expected to do. If governments make the slightest mistake and create a loss, then citizens react harshly and the government's power is threatened as happened in Iran in 1979. That is why they are risk averse.

Chapter 5 combines the theories of chapter 1 and chapter 4 to forecast oil prices in the future. According to the US Energy Information Administration (EIA) as well as other world oil supply forecasts, there will not be any world wide oil shortages for at least twenty years and possibly one hundred years depending on the assumptions. These forecasts however do not take into account OPEC risk aversion that will reduce oil supplies below the normal supply forecast. All of the forecasters assume that the world works in a similar fashion as the US does, which is as a set of risk neutral or even risk lover producers. In fact most of the world is risk averse, causing a much lower potential supply of oil for the world. Using a Hubbert curve with a risk averse scenario, it is easy to see that an oil supply shock will hit the world well before 2010 and probably before 2005.

This oil shock will be harsh. Unlike the oil shocks of 1973 and 1979, the next oil shock will have supply on a downward trend. Demand however, will have much less room to maneuver. In the oil shocks of the 1970's and 1980's, there were many and varied ways for the US and the world to reduce its use of oil such as switching electricity generation away from oil towards coal, nuclear and natural gas. There were also many industrial conservation methods available. But this time around, those easy reductions will not be available. Therefore, the next oil price shock could be a lot more devastating to our economy.

Chapter 6 explains the reason behind why oil is so much more valuable than any other energy resource. It explains how the history of energy uses corresponds with greater world economic achievement. Usually economists and the general public theorize that technology is the only reason that economic growth has been sustained for the last several millennia. However, another cause of growth is the type of energy resources used. Humans by using more high quality high-grade energy resources have been able to exploit economies of scale. In other words, it is not only technology that has created world economic growth, but rather the type of energy resources used. Those higher grade energy resources created growth through potential economies of scale.

For example, when humans invented fire, they created a technology that could keep them warm, provide for cooking, and provide light. However, before there can be a fire, there must be wood or a fuel source. The fuel is the natural resource. The fire is the technology. But to have it all work, both are needed. Wood as a fuel source though is a very high quality natural resource better then say grass. So it is not just technology by itself, but technology and a high quality, high-grade natural resource that makes the technology so successful. Fires as a technology would not have been nearly as successful creating economic growth if only grass could be used as a fuel source. Wood made fire successful.

Chapter 7 explains an interesting phenomenon called entropy subsidies. It is always the case that as one energy price gets more expensive, other alternative energy resources and energy technologies also become more expensive. The cause of this is the loss of entropy subsidies. For example, it can take one person one day to clear away a quarter of an acre of forest using a chain saw and a truck. However, without the availability of oil that same person might take a week to clear the same forest with an axe and a horse and buggy. Therefore with less high-grade energy resources such as oil, lower grade energy resources such as wood become more difficult and expensive to obtain.

This is true for many other technologies as well. The example given in this chapter shows how oil shale was developed in Colorado in the US. The interesting phenomenon is how as oil prices kept going up, so did the costs of providing oil from oil shale. The end result was that oil from oil shale was always expected to be cost effective but it never was. The reason for this was the loss of oil. As the economy had and used less oil, the price of that oil had to go up creating the oil shale cost spiral. A lack of oil made the oil alternative more expensive. This is a loss of an entropy subsidy.

Chapter 8 explains the chain of use of any and all energy resources, which is called the energy utilization chain (EUC). The EUC is a way of explaining how any given energy resource fits into an economy. Each energy resource must follow a chain of use where it is discovered, extracted, converted into a useful form, transported to where it is needed, and finally used in such a way as to give society an energy service. Each one of these links in the chain of the energy resource's ultimate use can create problems and extra costs for society. This is useful for discovering the weakest link, or more to the point, the most expensive link in the chain of energy use. It is useful to understand these links for energy alternatives. If society wants to know how successful alternative energy resources will be, we must look at how successfully an alternative will fit into the EUC. For example, solar energy may be free and abundant, but finding a way to store it may be a difficult link in its use. The EUC reveals these weak links. It shows where high costs and difficulties in the implementation of alternative energy technology emerge.

Chapter 9 analyzes all possible alternative energy resources using the EUC. It shows that once there is an oil shortage, alternative energy resources will be very difficult to use and therefore, very expensive. All of the most commonly regarded alternative energy technologies such as solar energy, wind energy and fusion power are analyzed using the EUC. This shows exactly where the limitations of each alternative energy resource are.

If all the alternative EUC's are analyzed carefully, the most logical conclusion you can make is that society has to change its lifestyle in order to be able to transition away from oil to any other energy resource. While changing our lifestyle is good news to most ardent environmentalists, it will not be easy. I have yet to see the kinds of sacrifice necessary to be able to implement such a change in lifestyles. I do not see anyone voluntarily moving to new lower energy use lifestyles especially in regards to oil use. This bodes very badly for the so-called "lifestyles EUC" explained in these chapters, where society is able to substitute away from oil use by merely changing its lifestyle. Extreme changes are needed. I do not believe people are ready for the changes that are heading for us quickly on the horizon. There will be a titanic battle between trying to hold onto the past glorious oil dominated economy, and the new low oil lifestyles that are to come.

Chapter 10 is an analysis about energy substitutes and how alternative energy resources as well as capital and labor can substitute for any particular non-renewable energy resource. Stiglitz (1979) and Hartwick (1977) have theoretically proven that an essential finite natural resource input can last forever if there is enough capital input to substitute for the resource. The problem though is physics. This chapter shows how the Stiglitz and Hartwick articles violate the laws of physics by assuming a rate of substitution between capital and energy or between any resource and capital that is simply physically impossible to do and unsustainable.

It is interesting that just like scientists and the general public are often not concerned with taking into account economics in their analyses, so also are economists often not concerned with taking into account physics in their analyses. The result is sometimes misleading. That is what happened here. The original Stiglitz analysis took a simple model and extended it beyond what is allowable by physics, even though the mathematics itself is correct. What is amazing though is how such intricate mathematics can be completely disconnected from the practical world, and impossible, and yet by using so much complicated mathematics the model has the appearance of plausibility. It is even published in the top economics journal. Thus mathematics has become more important than simple generally understood theories even at the expense of being incorrect. This chapter shows how it is incorrect.

One important aspect of Scarcity and Growth, as has been stated, is technology. If technology is powerful enough, then it can over come any resource shortage. What chapters 6, 7, 8, 9, and 10 show is that technology is not as powerful as you would think for over coming a resource shortage. Most people believe technology is all-powerful. But it is not. It is clear that technology, like natural resource supplies themselves, is limited, and therefore human's are limited in their pursuit of economic growth such that sooner or later our society will reach a limit and have decline. Economic decline has happened to many civilizations before us, and it will happen again to us. But this should not be cause for hopelessness, because even though we are limited as human's, we are also resilient and able to overcome harsh circumstances. History itself shows this. The Roman's survived the decline and fall of the Roman Empire, the Mayans survived the decline and fall of the great Mayan empire, and we shall survive any decline we must endure. For the true strength of our humanness is our adaptability not our power.

Looking at Scarcity and Growth, we can either be optimistic or be hopeful. Hope or optimism. Optimism is the idea that we can overcome nature and tame it. It is like what Julian Simon's (1981) has wrote about our inventiveness, and how it subsumes all barriers to greater luxury. On the other hand, there is hope. Hope is inward, not outward. Hope is where we feel good about ourselves no matter what we accomplish or don't accomplish, or no matter what our material status is. When a challenge confronts us, and we have hope, it does not matter if we tame the lion or simply learn to live with its danger. We do not need to overcome nature. With hope, we can adapt to it and live within our means. Therefore, we should not look at technology or resources as a means to greater luxury, but simply as a means to provide what we need. And as fewer resources are available, hope allows us to adapt to that. Hope does not go up and down with economic well-being, only optimism does that.

Chapter 11 is an interesting analysis of the fall of the Soviet Union. It gives an alternative explanation for the fall, which is that there was an oil crisis.

Most analysts believe that the fall of the Soviet Union was due to a lack of technological innovation within a corrupt and non-incentive giving system. However, the evidence overwhelmingly suggests that indeed there were incentives within the system not only to find oil but also to produce all other goods and to develop new technology. The incentives included bonuses and promotions within companies or within the communist party itself. Many good technologies were developed in the Soviet Union including the atomic bomb, the first space satellite, and the first manned space orbit. Therefore, there was never a dearth of new technology within the Soviet Union and there were great achievements. The Soviets had a robust economy.

Considering the wealth of innovation that the Soviet Union did manage to produce and the fact that the Soviets were so powerful militarily for seven decades, there must be an alternative explanation behind the fall. One alternative and obvious reason is that the Soviet Union suffered an oil crisis that caused an economic collapse. The Soviet's had a bigger oil production decline then the West ever experienced in the 1970's and it completely ruined the Soviet economy. In this chapter is shown all the evidence for this oil crisis that caused their economic collapse. This shows that a resource shortage is more powerful than politics or technology and that what happened to the Soviet Union will surely happen to us.

REFERENCES

Barnett, Harold J. & Chandler Morse.(1963). Scarcity and Growth; The Economics of Natural Resource Availability. Johns Hopkins University Press for Resources for The Future, Baltimore, pp. 4, 5, 6.

Barnett, Harold J. (1979). "Scarcity and Growth Revisited" in Scarcity and Growth Reconsidered, V. Kerry Smith Ed., Resources for the Future, Baltimore, pp. 163 – 217, p. 164.

Hartwick, J.M. (1977). "Intergenerational Equity and the Investing of Rents from Exhaustible Resources," American Economic Review, Volume 67, no. 5, pp. 972 –974.

Hotelling, Harold, 1931. The Economics of Exhaustible Resources. Journal of Political Economy. 39(2): 137 – 175, pp. 144, 152, 174.

Hubbert, M.K., 1962. Energy Resources, A Report to the Committee on Natural Resources: National Academy of Sciences, National Research Council, Publication 1000-D. Washington, D.C..

Menard, H.W. and George Sharman (1975). "Scientific uses of Random Drilling Models" Science, Volume 190, Number 4212, pp. 337 - 343.

Simon, Julian L. (1981). The Ultimate Resource, Princeton University Press.

Stiglitz, J.E. (1979). "A Neoclassical Analysis of the Economics of Natural Resources" in Scarcity and Growth Reconsidered, V. Kerry Smith Ed., Resources for the Future, Baltimore, pp. 36 – 66.

Chapter 1: The Mineral Economy: How Prices and Costs Can Falsely Signal Decreasing Scarcity*

ABSTRACT

Natural Resource prices and costs of extraction have declined simultaneously with increasing quantities of extraction for a long time. In a Hotelling sense this indicates decreasing scarcity since low cost resources normally would be used first and quantities of extraction normally would decrease over time. The main reason for the trend being opposite to Hotelling characteristics is usually thought to be due to technological innovation. However, an alternative reason for decreasing costs and prices and increasing quantities of extraction may be due to Georgescu-Roegan's (1972) concept of "Bonanza" where there is only the appearance of decreasing scarcity. Norgaard's (1990) "Mayflower" problem can be used to model an alternative neo-classical approach to resource extraction and scarcity. In this paper, a model of resource exploration is developed where the explorer does not know total reserves of the resource base as he searches for and extracts the natural resource. The explorer never entirely knows how big the resource base is but does gain information about the potential location of new reserves as discovery proceeds. That reduces exploration costs.

---

* Reprinted from *Ecological Economics*, Volume 31, Number 1, Douglas B. Reynolds, "The Mineral Economy: How Prices and Costs Can Falsely Signal Decreasing Scarcity," pp. 155-166,

The lower exploration costs can cause the price to fall over time, until eventually scarcity of the resource causes the price to rise. The true scarcity is only revealed towards the end of exhaustion. The model shows that it is possible to have several years of increasing production simultaneous with lower prices and costs until a sudden, intense price rise occurs with a huge cut in production. When technology is able to cut costs and increase the reserve base, the decline in prices and costs and the increase in production can last longer. However even with better technology, it is still possible for a sharp increase in price as long as demand is growing faster than technological innovation. The problem is that the true size of the resource base is never known. Society does not know if technology is actually overcoming scarcity or not until demand for a resource outstrips supplies. It is even possible for a price shock of incredible magnitude to surprise an economy within one or two years after a hundred years of declining prices and increasing production.

INTRODUCTION

According to Hotelling (1931) there are 3 things, we can say about resource extraction for a finite resource. 1) Firms will always try to extract low cost resources before high cost resources causing resource extraction costs to increase over time. 2) In general, greater scarcity caused of in situ reserves increases the value of a resource causing the price to increase over time. 3) Since the price of a resource increases over time, then demand and production should decrease over time. Barnett and Morse (1964), and more lately Scott and Pearse (1992) and Simon (1996), show that all of these theoretical trends have not occurred. Costs of extraction and prices of natural resources have mostly decreased over time, not just in recent years but in many cases for decades. Quantities extracted and produced of most natural resources have also increased over the same long time frame. The only logical explanation is that technology and substitutes are in fact more powerful than scarcity.

In this paper is shown an alternative model for scarcity based on Norgaard's (1990) Mayflower problem that shows how it is possible for there to appear to be long trends of decreasing scarcity only to have a sudden and dramatic increase in scarcity. The model here shows that long-run resource exploration cost, and resource price declines simultaneous with increasing resource extraction, do not necessarily signify decreasing scarcity. The paper shows that the true power of technology in over coming scarcity may be difficult to determine. It is also shown how price and cost measures of scarcity may fail even when prices and costs have declined for many years.

Hall and Hall (1984) define scarcity in terms of the cost of extracting a resource, which is based on Hotelling's original concepts. Malthusian scarcity is where in situ stocks are finite and Ricardian scarcity is where stocks are not finite, but costs of extracting them increase as you extract more. Concentrating on Malthusian scarcity, Hall and Hall define Malthusian Stock Scarcity (MSS) where resource stocks are finite but are of uniform quality so that the cost of producing them stays relatively constant with constant technology. However instead of looking at the cost of extraction to determine scarcity, this paper will look purely at the costs of exploration. Extraction costs will be zero. The assumption of zero extraction costs can be modified later, nevertheless this assumption changes the emphasis of determining scarcity away from the costs of producing a resource and toward the cost of finding a resource.

Norgaard looks at the difficulty of exploring for resources in the "Mayflower" problem where information about the location of low cost natural resources is incomplete. When the Mayflower pilgrims came to America, they settled at the first most convenient location possible in New England even though the soil there was not as good as the soil in say the Midwest. If the Mayflower pilgrims had used the available resources optimally in a Hotelling sense, they would have taken the Mayflower to the Gulf of Mexico and up the Mississippi River all the way to Iowa in order to start their farms there. Or at least within a generation or two, the new settlers would have moved all their settlements to the

Midwest. At least part of the reason that the Mayflower pilgrims and subsequent generations did not go to the Midwest quickly was because they did not know that there was good soil there. Their actions imply that finding information about resource location is an unavoidable cost of the supply of natural resources.

MONTE CARLO MODELING

In order to understand the economics of mineral exploration and the Mayflower problem, consider a Monte Carlo simulation. Menard and Sharman (1975) show in a Monte Carlo model that if oil explorers in the US drilled random oil wells throughout the US starting in 1859, then the super giant East Texas oil field would have been discovered at the very least by 1902, given the historical rate of exploration. East Texas was actually discovered in 1930. Other giant and super giant fields also undoubtedly would have been found earlier by merely drilling exploration oil wells randomly. The problem with the analysis is that Menard and Sharman know something that the early oil explorers did not know, namely that the East Texas oil field exists. In other words, the Monte Carlo modelers have perfect information that twenty-two giant oil fields in the US exist before they run their model, but the early oil explorers did not have this information and did not know if even one giant oil field existed until the first one was found. Therefore, consider an alternative set up.

A Monte Carlo model is rather like playing roulette where there are black and red slots on the roulette wheel for the ball to fall into. Black is oil, red is dry. The probability of getting black on a bet that costs the amount of money for drilling an oil well depends on how many black slots there are as a proportion of all the slots. If you were to play the gamble, the expected results are exactly equal to the probability of getting black. However, different roulette wheels like different countries and different regions have different probabilities of success. For example compare finding oil in the Persian Gulf to that of Switzerland. In other words, there are different roulette tables with different probabilities.

Assume then that there is a large room full of roulette tables, and that gamblers will come into the room to play the tables. Based on their expected probability of winnings, they will be more willing to bet on the tables with the higher probability of black since those tables will have a higher expected gain. Now assume one more thing. All the roulette tables are covered and mixed before the gamblers enter the room. In that case, the gamblers do not know which tables have a high proportion of black slots and which have a high proportion of red slots. They do not even know if any tables have any black slots. The gamblers have no idea what their chances of winning are or even if there is anything worth betting on or not.

As the gamblers come into the room with the covered roulette tables, they must try to get information about what the tables are like. What strategy will they use? Undoubtedly, the gamblers will wait to see what other gamblers do. If one gambler does make a bet and does get a black slot, chances are good that at least that particular roulette table has more black slots. The gamblers race to that table and start betting. However as the bets are made, every time a slot is hit, it is removed and not replaced. Eventually the black slots are used up and the gamblers will have to leave old tables and start taking their chances on newer tables.

The gamblers do not randomly play on all the roulette tables, as implied by a pure Monte Carlo model. For all they know, all the tables could have only red. There is no way of knowing a priori what the probability is that any black slots exist. In this situation, the gamblers have in their mind a subjective probability of winning at any given table. However, their subjective probability is different than the actual probability of the table. When the gambler goes to a table where previous gamblers won black, he has information to judge the probability that he also will win based on past experience. That is his own subjective probability. It is based on available information. A gamblers subjective probability will be higher at a table with a past history of winning bets than with an unknown table where there is no history. The subjective probability of getting black at a new table may be lower or even much lower than the actual probability of getting black at that table. Therefore there is a difference between a gamblers or a mineral explorers subjective probability of finding a mineral and the actual probability of finding a mineral, should he drill a well.

This is the Mayflower problem. At first there is very little information and few bets are made, i.e. explorations for minerals. Then as more information is found, more betting occurs because the information helps increase the probability of success until eventually scarcity brings to an end any chance of finding new resources. The mineral economy below is derived from this Mayflower problem. By using this model, we get a very different perspective on how a resource dependent economy will act over time and on the interpretation of the effect of better technology on scarcity.

THE MINERAL ECONOMY MODEL

Two economic models describe a resource-based economy in its purest form: the "corn" economy and the "hardtack" economy explained below. For a complete analysis of these models see Page (1979). We can call a third type of resource model the "mineral" economy. The mineral economy model is similar to a hardtack economy model except that the mineral – the hardtack – is hidden. The hidden hardtack must be explored for and extracted before it can be used. To explain the models, consider Robinson Crusoe stranded on a deserted island. In the corn economy scenario, he has some corn with him. He can choose to use varying amounts of the corn to either plant or eat depending on what his intertemporal preference is. If he eats more corn now, he has less corn to plant for future output. On the other hand if he plants more corn, he will have more future output but less to eat right now. In an infinite horizon world, such as a tribe of constant population on the island, a corn economy can last indefinitely.

In the second scenario, Crusoe is faced with a much more sobering possibility of only having a finite amount of hardtack. Again, depending on his personal preference, his years of life and the amount of hardtack he has, he faces several scenarios for exhausting that hardtack. He can choose to eat more now or save his stock and eat more later, but his reserves are limited. He may even not survive long if his supply is too small. If we extend Crusoe's life indefinitely by replacing him with a tribe on the island with a constant population and still a finite amount of hardtack, then that society faces the inevitable dilemma of sooner or later running out of hardtack and dying off. Better technology, though, could over come such a grave scenario.

However, neither of these scenarios for Robinson Crusoe parallels the real world, where our economies are to a large degree based on underground non-renewable non-recyclable resources, such as oil, coal and natural gas. In the real world, unlike a hardtack economy, these natural resources are not stored in such a manner so that we can quantify them costlessly, the way we can quantify hardtack stores costlessly. Rather, many natural resources are hidden from the economy until they are found. The hidden resources create an unavoidable cost for finding the minerals. That is an exploration cost. Even if extracting the mineral is costless, there is still an exploration cost implied by the search process that influences the rate of extraction. Using the characteristics of the Mayflower problem, we can create a model for the cost of exploration over time and then use that model to analyze scarcity and how powerful technology really is for overcoming scarcity.

Consider a situation where Robinson Crusoe lands on a deserted island with no food. Assume that previous shipwrecks with no survivors have left hardtack on the island. If Crusoe is able to find these old shipwrecks and the hardtack on them, he can survive by eating the lost hardtack. However, let us further assume that these previous shipwrecks happened many years before Crusoe's shipwreck, such that all the ships have been buried deep in the sand and are not visible from the surface. Crusoe can survive by eating this leftover hardtack, which is his "mineral" resource, if Crusoe can find it. This situation is the mineral economy. In the mineral economy assume there is Malthusian Stock Scarcity where the resources are of constant quality and even free to extract. The only problem is knowing how much resources there are and finding them.

When Crusoe first lands on the island, he must hurry to find food where he can. Because no previous shipwrecks can be seen from above ground, Crusoe does not know if any hard tack even exists underground and if it does exist, how much there is. Nevertheless with no other option, Crusoe takes a chance and digs for hidden hardtack. Eventually he gets a break and finds one shipwreck and manages to find the left over hardtack of that ship. Such a find will give Crusoe new information about where other shipwrecks may be because often times ships will wreck at the same location. This is similar to all the gamblers with the covered roulette wheels all going to the one table where there was a black slot found and betting on it. Therefore, with this new information about one ship's location, Crusoe will be better able to find other ships and his cost of searching will decrease. The more hardtack that Crusoe finds, the more information he has, and the easier it is to find even more hardtack.

Eventually though, because the supply of hardtack is finite, depletion sets in and Crusoe finds less and less hardtack. Therefore, even though Crusoe has more and more information about where the hardtack is, the depletion of the resource base creates higher costs of exploration, since it is hard to find the last few remaining stocks the rate of discovery decreases. Crusoe's supply of hardtack left to be found dwindles until he can no longer survive.

Two long-run forces, then, drive Crusoe's mineral economy. First, there is the information effect where the more information Crusoe has about the location of the hardtack, the cheaper it is to find and the more he is able to find. Second, there is the depletion effect where the less hardtack that remains to be found, the costlier it is to find and the less he can find. These forces create a pattern of first an increasing and then a decreasing rate of discovery. Uhler (1976) calls the two opposite forces of information and depletion the information and depletion effects.

AN EXPLORATION INFORMATION COST MODEL

Assume that Crusoe's technology for finding and using the hardtack stays constant and actual extraction is costless. Changes to these assumptions are possible after the basic model is developed. Assume every time Crusoe explores a location for hardtack, he must eat h_c of hardtack in order to give himself the physical energy to look. This is his cost of exploration effort. If Crusoe searches in 20 locations, he must eat 20 times h_c hardtack. Furthermore, assume for simplicity that when hardtack is discovered in any one location, that it is always the same amount of hardtack, call it h_d. While most minerals occur in large and small loads or pools, rather than constant loads, it is assumed that a large load is the same as a lot of small ones that are near each other. Therefore, hardtack that is left in the ground from previous shipwrecks is always packed in small tins or boxes that are the same h_d size. The tins are separate and must be discovered separately. Smith and Paddock (1984) suggest based on empirical evidence that minerals such as oil occur in large and small pools or loads and that the largest pools or loads of minerals are usually found first and then subsequently smaller pools and loads are found within one region. However, their empirical evidence is not general enough, since in many cases the largest loads of minerals are not the very first reserves found. For example, the East Texas oil field, Prudhoe Bay, and the great Ghwar oil field in Saudi Arabia were not the first oil fields found even though these are some of the largest fields on Earth. In general, small loads of minerals are found first, then large ones and then small ones again. For Crusoe, it is assumed that each tin is separate and equal, however large loads of minerals can be simulated by having lots of tins being close together. When one is found others nearby are easy to find. Hence:

h_c = Pounds of hardtack consumed per exploration location that is searched, i.e. effort cost.

h_d = Pounds of hardtack found per exploration location when the exploration is successful. h_d is the pounds of hardtack found in each tin. All tins are assumed constant for all successful exploration locations.

Where $h_d > h_c$.

In order to discover Crusoe's marginal cost of extraction for any given rate of extraction, it is necessary to know the probability of success for each individual search and how that probability changes as the rate of extraction changes. Crusoe's exploration of any given location is not always successful. There is a probability of success or failure. The probability for Crusoe's success rate is called the probability weight function, P_r, which is modeled formally below. P_r is a function of ultimately recoverable reserves, the current rate of exploration, and information, which is dependent on cumulative discovery. P_r models Crusoe's subjective probability of finding more hardtack rather like a probability coin. Every time Crusoe expends effort to search for hardtack at a specific location, his success of finding more hardtack is like flipping a coin. However, the probability of success is not necessarily 50/50. Instead, the probability weight of the coin can be anywhere from 0/100 to 100/0 for success, depending on Crusoe's subjective probability. That subjective probability depends on information such as how successful previous explorations have been and exploration progress. The subjective marginal benefit of exploration for Crusoe is then $h_d \cdot P_r$ or alternatively the marginal cost is h_c / P_r.

In the Monte Carlo game with the gamblers, there was the actual probability of taking a gamble on the roulette wheel, which the gamblers did not know, and their own subjective probability of gambling at that table. Like the gamblers in an unknown Monte Carlo simulation, Crusoe too has no idea if there is any hard tack at all under ground or, if there is some, where it is. Nevertheless, Crusoe will use whatever information he can garner from past experience to choose the best location possible. His subjective probability, P_r, will change as information changes.

There are two distinct aspects of how information affects Crusoe's exploration process. When Crusoe begins his search for hardtack at the beginning of the day, he will have in his mind all possible information about where hardtack may or may not be. He will begin his search at the location where his subjective probability of success is the highest. That is, Crusoe will explore for hardtack first in the most probable locations. He will continue searching at second and third best locations. Therefore, as the day passes, Crusoe will have a lower and lower probability of success.

During the day, however, Crusoe may also gain new information, which could help his search. For example, Crusoe may discover hardtack and start to see a pattern about where hardtack usually occurs underground. This may increase his probability of success. Therefore, there are two distinct forces that affect the probability of success. One force is where the old information becomes less and less useful in helping Crusoe make successful new discoveries. The other force is where new information discovered during the course of exploration helps Crusoe raise his probability of success. In order to understand these two forces better let us separate them and model them differently. For simplicity, assume that Crusoe can only use one set of information during a given day and that any new information he finds he cannot assimilate until the next day. This will separate the effects of information into a short run (within a day) and a long-run (longer than one day) time frame. In other words, Crusoe will use information rather like capital. "Information capital" is Crusoe's accumulated information, i.e. his stock of information. Information capital is like a stock of capital machinery, it cannot change in the short run. Only in the long-run can new information be added to the current stock of information, causing the information capital to increase.

SHORT RUN CHARACTERISTICS OF PROBABILITY

In the short run, Crusoe has a finite stock of information, i.e. information capital that cannot change, to use for looking for more hardtack. The information capital is a complex data configuration. New information and data cannot be easily integrated into the existing information capital just like new machinery cannot be easily and quickly integrated into an existing stock of capital. Therefore, all new information cannot be assessed until the next day. However, Crusoe can use the existing information capital to look for hardtack over an infinite combination of places during the current day. He will look for hardtack first in the places where his subjective probability of successfully finding hardtack is the highest and then in second and third best locations. As he looks, though, the probability of success declines. The reason for the decline is because the information capital is subject to the law of diminishing returns, which causes the value of that information stock to decrease. In other words, the law of diminishing returns degrades the value of Crusoe's information capital as he uses it so that it is less and less useful the more it is used during the day. The marginal productivity of the information capital declines as effort and therefore output increases. This is similar to how the marginal productivity of capital itself declines as output increases. The effect of the degrading value of old information is to cause a decreasing probability of success. A convenient formula for P_r to model this short run value of the information capital is the following.

$$P_r = \exp[- Q/L]$$

$$Q = Q(S)$$

Where $dQ/dS > 0$; and $Q =$ the daily quantity discovered or current rate of discovery; $S =$ effort, in terms of numbers of searches per day; $L =$ a long-run trend parameter.

The short run probability will act as a decline curve as the rate of discovery increases. The rate of effort, S, is an independent variable chosen by Crusoe, which will determine the rate of discovery. If effort increases, discovery increases and probability falls. At the beginning of the day it may take two exploration tries to find hardtack, but as the day wears on, it may take Crusoe ten tries. Therefore, the probability of finding that last tin of hardtack decreases from 50% to 10%, and hence the marginal cost rises. The lower probability of success is due to the diminishing returns on the current information capital, which increases marginal cost. Crusoe will tend to look for hardtack during the day until his cost of exploration is close to or equal to his benefit of successful discovery, h_d. Setting the probability as a function of the rate of discovery rather than effort makes the model easier to work with. The more one discovers the resource during the day, the more one uses the current information capital. The more the current information capital is used, the less valuable it is. The less valuable the information capital is, the lower the probability of success. Therefore marginal cost rises.

LONG-RUN CHARACTERISTICS OF PROBABILITY

Recall in the Monte Carlo situation above that initial gambles, like initial efforts to find a resource, are very unsuccessful due to a lack of knowledge and therefore a great deal of effort is used to find only a small amount of resource. Eventually, however, with greater information, it takes fewer efforts to find a given amount of resource. Therefore, in the long-run, new information will increase the information capital and increase the probability of success, such that the probability of finding the mineral will increase. Concurrently, however, there is depletion. In addition to long-run information gains, the depletion effect will cause less of a resource to be found. As a resource depletes to nothing, the probability of finding more of that resource must also decline. Therefore, in the long-run, two effects on probability are modeled: the information effect and the depletion effect. See Uhler (1976).

With greater information capital Crusoe can find more hardtack for a given level of effort. Greater cumulative discovery implies greater information capital, which will cause a higher probability of success. This is the information effect. However, with greater depletion, Crusoe will find less hardtack with a given level of effort. Greater cumulative discovery equates to less remaining resources in the ground available to find, which will cause a lower probability of success. This is the depletion effect. The information effect will dominate initially, and the depletion effect will dominate subsequently. The probability weight function as a whole shifts first outward and upward and then inward and downward with cumulative discovery. The result is a small-large-small sequence of the discovery and extraction of hardtack in the long-run. These long-run effects on probability are modeled using cumulative discovery.

Greater cumulative discovery of hardtack implies that there have been greater exploration efforts, and such efforts must increase the amount of information. Therefore, cumulative discovery is used as the long-run information capital indicator. It is also possible to use cumulative exploration as an information indicator, but if unsuccessful search locations actually do provide information, then they will lead to more discoveries. Thus cumulative discovery is just as good an indicator of information capital as cumulative exploration efforts. The benefit of using cumulative discovery rather than cumulative effort is that one variable can be used for both the information and depletion effects more simply.

Greater information capital shifts the probability weight function higher. That shift is an information effect shift, where greater cumulative discovery <u>implicitly</u> signals greater information capital. As the stock of undiscovered reserves depletes to zero, though, the probability of finding new hardtack must eventually become zero, because there is no more hardtack left to find. The depletion effect then is where less hardtack reserves make it more difficult to find new hardtack causing the probability weight function to shift down. The depletion effect depends <u>explicitly</u> on greater cumulative discovery, because higher cumulative discovery equates to a lower stock of remaining undiscovered reserves.

In order to model the information and depletion effect, assume the long-run trend parameter, L, shifts the probability function P_r in the manner explained above. A likely equation for the trend is M. King Hubbert's life cycle hypothesis (Hubbert 1962). Hubbert's life cycle hypothesis, which was used to explain the long-run trend of oil production in the US, ideally models a mineral economy faced with the Mayflower problem. Hubbert's model has a small-large-small sequence of discovery and production of a mineral that would be expected in the mineral economy. For an economic analysis of Hubbert see Cleveland (1991), Cleveland and Kaufmann (1991), McCabe (1998), and Edwards (1997). Hubbert's original model using L as the trend for current discovery is a time dependent logistics curve as follows:

$$L = a \cdot U \cdot \{1 + \exp[-a(t-t_o)]\}^{-2} \cdot \exp[-a(t-t_o)] \qquad (1)$$

Where a = an amplitude parameter, (this determines the height and width of the Hubbert curve); t_o = the time at which the peak in production occurs; t = Time; U = Ultimately Recoverable Reserves; L = the trend of the rate of discovery.

However, it is possible to subsume the time element of Hubbert's logistics curve in order to make it a pure cumulative discovery model that can be used to measure the combined effects of information and depletion. Hubbert's cumulative discovery model is as follows,

$$CQD = U/\{1 + \exp[-a(t-t_o)]\}$$

Which can be rewritten as

$$(t-t_o) = \{Ln[U/CQD - 1]\}/ -a \qquad (2)$$

where CQD = Cumulative Discovery.

Substituting equation 2 into equation 1 gives

$$L = a \cdot CQD - a \cdot CQD^2/U \qquad (3)$$

which can be called the Hubbert trend.

J.J. Wiorkowski [1981] developed a range of logistics curves similar to Hubbert's life cycle curve, using what is called the Richards function (F.J. Richards [1959]). From Wiorkowski and Hubbert, a generalized trend in terms of cumulative discovery can be used to describe L over time. The generalized Hubbert trend is as follows.

$$L = a \cdot CQD \cdot U^{(\mu-1)}/\mu - a \cdot CQD^{\mu} \cdot CQD/U \cdot \mu$$

and, when $\mu = 0$,

$$L = a \cdot CQD \cdot \ln U/U - a \cdot \ln CQD \cdot CQD/U$$

Where μ = shape parameter, $-1 < \mu <$ infinity. Equation 3 was Hubbert's original trend where $\mu = 1$.

The Hubbert trend, L, drives the probability weight function in the prescribed manner hypothesized by the mineral economy. It models an early low probability of success that increases due to the information effect. It then models a subsequent decreasing probability of success due to the depletion effect. In the short run, L does not change so that as Q increases there are diminishing returns to finite information capital causing a decline curve for the probability weight function. The long-run trend will cause the short run probability function curve to shift out initially and then inward in later periods.

A MINERAL MARKET SIMULATION

Since Crusoe does not know how much hardtack is under the ground, it could be one million times his most opulent lifetime consumption or only a weeks worth of food, he cannot perform an intertemporal maximization. If Crusoe's discount rate is zero, he simply lives at a subsistence level. However, for any other discount rate above zero, he just does not have enough information to maximize utility. Nor can a tribe of constant population. As long as a tribe has a rate of discount greater than zero, then they cannot use the Hotelling principle, since they do not know if the resource base will last a thousand generations or less than one. Therefore, a simple Crusoc model cannot be solved. However, if we assume that Crusoe or a tribe believes that the resource is so vast that it verges on being infinite, i.e. that the probability of discovery always goes up in the long-run, then there would be no Hotelling rent. The resource is seemingly renewable. Crusoe will each day explore up to the point where marginal cost equals h_d. Consider this possibility more generally.

Assume instead of Crusoe by himself, that there is a large economy and that there is a mineral industry for one particular essential natural resource input into that economy. The market for the mineral is within a larger market place. The marginal cost of finding the mineral for all firms in the industry is:

$$MC = E_c/h_d \cdot P_r$$

Where E_c = cost of effort in the industry; MC = aggregated marginal costs of the industry; P_r = exp(-Q/L). P_r is determined by the industry wide rate of discovery.

When the industry extraction rate, Q, increases, P_r decreases due to industry wide diminishing returns to information capital. Note, the industry is made up of many firms and all firms use each others information, therefore, this is an efficient market where all information is available to the other players, similar to the Monte Carlo gambling simulation above. Costs are minimized, profits maximized. h_d is the marginal unit of extraction, therefore let h_d = 1 for convenience. The cost of actual extraction is assumed zero. That assumption can be changed to add an extraction cost that also decreases with time, but the general conclusion will be the same. Therefore in this simulation it is assumed that the aggregated marginal cost includes both extraction and exploration costs. In that case:

$$MC = E_c \cdot \exp(Q/L) \tag{4}$$

This gives the following characteristics:

$dMC/dCQD < 0$ when the information effect dominates

and $dMC/dCQD > 0$ when the depletion effect dominates

In addition technological effects for supply can be added to equation 3 and 4. Let:

$$E_c = E_{co} \cdot \exp(-T_c \cdot t)$$

Where E_{co} = Initial cost of effort in the industry; T_c = Cost technology factor (the percent reduction in costs per year). Also

$$U = U_o \cdot \exp(T_u \cdot t)$$

Where U_o = Initial Ultimately Recoverable Reserves; T_u = Ultimately Recoverable Reserve technology factor (the percent increase in ultimately recoverable reserves per year due to technology)

Consider a simple demand curve for the resource that reflects a growing economy.

$$Q = G \cdot (P)^{-1/3} \tag{5}$$

and $$G = G_o \cdot \exp(g \cdot t)$$

Where G = economic production of the entire economy; G_o = initial economic production of the economy; g = economic growth factor for the economy; P = price for a given quantity demanded.

According to the Hotelling rule, rent will increase over time. However, if uncertainty is large enough this may not happen. See Watkins (1992). For example, if resource prices have been falling for a very long time, and they look to continue to fall for a long time, then producers will set P = MC. There will be no Hotelling rent. Producers believe the resource is infinite due to new technology. When technology is thought to be the cause of a long price decline, then it is easy to conclude that technology will continue to improve indefinitely and cause prices to fall indefinitely. In that case, a resource producer maximizes present value not by withholding exploration for, or extraction of, resources, but by producing more now before prices go even lower. The producer maximizes present value simply by producing as much as profitable in the current year irrespective of rent. Producers would generally set P = MC.

Consider a model simulation using the following values for equations 3, 4, and 5: U_o = 100,000; a – 0.15; T_u = 1% per year, i.e. slower than economic growth; G_0 = 500; g = 3% per year, i.e. faster than the rate of technological growth of the resource; E_{c0} = 1; E_{c13} = 1, E_c is held constant in the initial 13 periods to give the system stability; T_c = 5% per year after the first thirteen years.

In the simulation, the technology to increase ultimately recoverable reserves is improving at a slower pace than economic growth. However, the effect of technology on the resource base to increase its recoverable reserves is unknown to the market and remains unknown until late in the exploration process. Therefore the market sets P = MC. In each period, Q and P are solved using equations 3, 4 and 5, then the effects of technology are added. Since equation 4 and 5 are indeterminate for Q, Q_{t-1} is used to solve for Q_t. The difference between Q_t and Q_{t-1} is small enough so that the substitution creates no problem.

Figure 1.1. Price, Production, and Hubbert Supply Trend as a Function of Cumulative Production

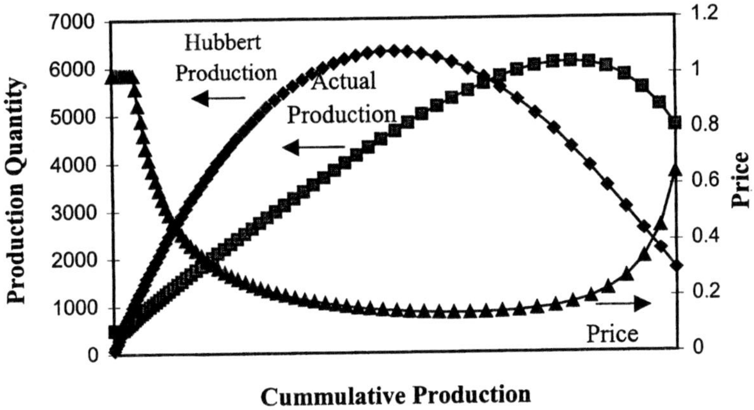

The results of the simulation can be seen in Figure 1.1 and 1.2. In Figure 1.1, the supply Hubbert trend (L), the quantity produced (Q) (which is the equilibrium of supply and demand), and price (P) are shown as a function of cumulative production (CQD). Actual production is less than the supply trend potential, i.e. the Hubbert trend, for most of the cycle. It is only after the supply Hubbert trend reaches its point of decrease that production finally gets close to and goes above the Hubbert trend. The Hubbert supply trend is only the *trend* of supply, it does not determine absolute supply. The actual supply can be above or below the Hubbert trend although supply becomes increasingly inelastic above the Hubbert trend. Scarcity is revealed only when production goes well above the Hubbert trend. At that point, the price finally starts increasing and production finally starts decreasing. Before that point, it is not possible to know how forceful technology is for increasing potential reserves. Suddenly, scarcity is revealed but unlike the Hotelling model, there is no forewarning.

In Figure 1.2 is shown the price trend as a function of time. Notice that the price, as well as the costs that determine price, which is not shown, decline for many years before the price finally increases. Furthermore, during those price decreases, quantities extracted of the resource are constantly increasing. In fact it is possible to model a situation where price decreases and supply quantity increases continue for hundreds or even thousands of years before the price finally rises and quantities finally decline. However, the change from one trend to the other can happen within a few years even though the trends before the change over lasted for hundreds of years. In other words, the information effect can dominate the depletion effect for a very long time before depletion finally and dramatically dominates. If you add to the information effect technology, then it takes even longer before depletion dominates. Nevertheless, the power of technology to overcome scarcity is not known until it is revealed late in the cycle.

Figure 1.2. Price and Production as a Function of Time

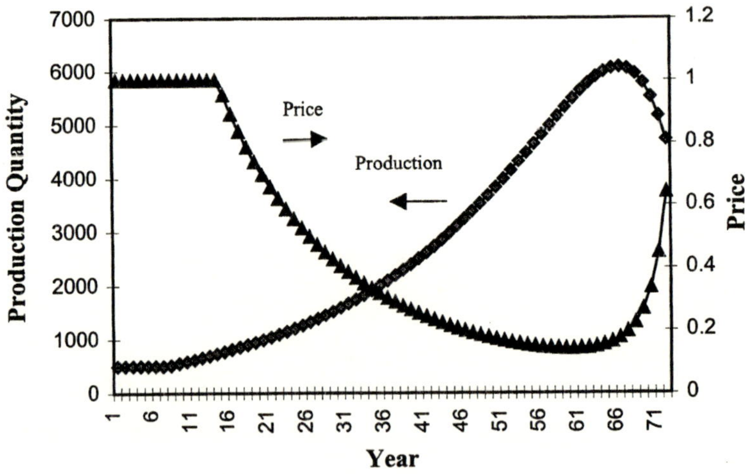

The problem with using price, cost or quantities produced as a measure of scarcity is that up until the point where demand crosses the Hubbert trend, it is unclear how powerful technology is for decreasing scarcity. Only when prices start actually rising due to the depletion effect can society know how big the Hubbert cycle is and how much technology has stretched the resource base. Before that point in time, the information effect obscures the size of the resource base and the true power of technology for reducing scarcity. An additional problem is that since price declines for a long period of time before the price shock, it is unclear what substitutes for the resource are available and at what price. There has been no gradual increasing price and therefore no signal to inventors of when a backstop will be needed, and no signal to consumers of when the backstop will be needed to be used. The market is not properly prepared for the change over.

However, once the price reaches its lowest level in Figure 1.2, producers of the resource will realize that the resource is a lot scarcer then they had previously thought. They will suddenly understand that the price will start increasing and they will cut production in order to maximize intertemporal profit via the Hotelling principle. This cut in production will cause the price shock to be much more sever. In addition, Reynolds (1997) shows that the backstop price of a resource can unexpectedly rise due to the loss of entropy subsidies. It is conceivable that after one hundred years of price decline and production increase, that a resource can have a ten or one hundred fold price increase within a year or two, with a corresponding decline in production. Given that short run elasticities of demand and substitution are much more inelastic, then long-run elasticities, this can create quite a problem for an economy to say the least.

CONCLUSION

In the Hotelling model, there is perfect information about the backstop price, the quantity of the resource, the demand for the resource, and the cost of extracting the resource. With perfect knowledge, rent of a resource increases at the interest rate up to the backstop price. Uncertainty theory can be added to the Hotelling model to try to find out how trends of prices and quantities change under various assumptions when the resource base is unknown. However if there is enough uncertainty it is almost impossible to use the Hotelling principle. In this model is shown how the uncertainty of the resource base can obscure the actual trend of scarcity and the true power of technology and create a price shock. Such a price shock can occur after a very long price and cost decrease simultaneous to a long production increase. This model shows theoretically that empirical data on cost and price declines as well as data on increases in quantities of extraction do not necessarily imply decreasing scarcity. Indeed, such data may not even be useful for warning about impending price shocks. What is even more foreboding is that once producers do know that scarcity is much greater than previously thought, the price shock will become all the more severe. Resource producers will cut back production due to increasing prices causing prices to increase even further. The so called back stop price may be much higher than expected especially since the loss of entropy subsidies can make the backstop price increase substantially.

REFERENCES

Barnett, Harold J. & Chandler Morse.1963. Scarcity and Growth; The Economics of Natural Resource Availability. Johns Hopkins University Press for Resources for The Futurc, Baltimore.

Cleveland, Cutler J., 1991. Physical and Economic Aspects of Resource Quality, The Cost of Oil Supply in the lower 48 United States, 1936 – 1988. Resources and Energy. 13: 163 - 188.

______________ and Robert K. Kaufmann, 1991. Forecasting Ultimate Oil Resources and its Rate of Production: Incorporating Economic Forces in the Models of M. King Hubbert. The Energy Journal. 12(2): 17 - 46.

Edwards, John D., 1997. Crude Oil and Alternative Energy Production Forecasts for the Twenty-First Century: The End of the Hydrocarbon Era. American Association of Petroleum Geologists Bulletin. August 81(8): 1292-1305.

Georgescu-Roegen, Nicholas, 1972. Energy and Economic Myth. in: Nicholas Georgescu-Roegen, Energy and Economic Myths: Institutional and Analytical Economic Essays. Published 1976, Pergaman Press, New York, 3 - 36.

Hall, Darwin C., and Jane V. Hall, 1984. Concepts and Measures of Natural Resource Scarcity with a Summary of Recent Trends. Journal of Environmental Economics and Management, 11: 363 - 379.

Hotelling, Harold, 1931. The Economics of Exhaustible Resources. Journal of Political Economy. 39(2): 137 - 175.

Hubbert, M.K., 1962. Energy Resources, A Report to the Committee on Natural Resources: National Academy of Sciences, National Research Council, Publication 1000-D. Washington, D.C., 54, 61, 67.

Menard, H.W. and George Sharman. 1975. Scientific uses of Random Drilling Models. Science. 190(42): 337 - 343.

McCabe, Peter. 1998. Energy Resources-Cornucopia or Empty Barrel?. American Association of Petroleum Geologists Bulletin. November 82(11): 2110 – 2134.

Norgaard, R.B., 1990. Economic Indicators of Resource Scarcity: A Critical Essay. Journal of Environmental Economics and Management. 19(1): 19 - 25.

Page, Talbot, 1979. Conservation and Economic Efficiency. The Johns Hopkins University Press, for Resources for the Future, Baltimore.

Reynolds, Douglas B., 1998. Entropy Subsidies. Energy Policy. February, 26(2): 113-118.

Richards, F.J., 1959. A Flexible Growth Curve for Empirical Use. Journal of Experimental Botany. 10: 290 - 300.

Scott, Anthony and Peter Pearse, 1992. Natural Resources in a high-tech Economy: Scarcity Versus Resourcefulness. Resources policy. 18(3): 154-166.

Simon, Julian L., 1996. The Ultimate Resource 2. Princeton University Press, Princeton.

Smith, James L. and James L. Paddock, 1984. Regional Modeling of Oil Discovery and Production. Energy Economics 6(1): 5 - 13.

Uhler, R.S., 1976. Costs and Supply in Petroleum Exploration: The Case of Alberta. Canadian Journal of Economics. 9(1): 72 - 90.

Watkins, C. J., 1992. The Hotelling Principle: Auto bahn or Cul de Sac?. The Energy Journal. 13(1): 1 - 24.

Wiorkowski, J.J., 1981. Estimating volumes of remaining fossil fuel resources: a critical review. Journal of the American Statistical Association. 76(375) 534 -547.

Chapter 2: Determining the Power of Prices to Change Oil Discovery and Production Using a Non-Time Dependent Hubbert Model*

ABSTRACT

M. King Hubbert's original oil production and discovery forecasting model (Hubbert 1962) used to determine US ultimately recoverable oil reserves is now well known in the literature. This paper will build on Hubbert's model using a non-time dependent variation of his curve called the quadratic Hubbert curve, which can be used more easily for policy analysis. A quadratic Hubbert curve can be used as a "Black Box" for policy makers in order to determine any particular region's potential supply of oil. The paper will test how powerful price is for increasing or decreasing oil discovery as well as production above or below the Hubbert curve. The US lower 48 oil-producing region will be tested since it is a mature oil-producing region with one of the most competitive markets in the world. The results show that price is not a powerful determinant of supply, that is supply is very inelastic with respect to price. Not only is the supply of production inelastic, but the rate of discovery is also inelastic with respect to price.

* "Determining the Power of Prices to Change Oil Discovery and Production Using a Non-Time Dependent Hubbert Model," by Douglas B. Reynolds, working paper, under review at *Ecological Economics*, printed with their permission.

INTRODUCTION

Determining the supply of oil for a given country or region has been an important issue for some time. One of the most highly regarded methods of determining oil supply is the logistics model that M. King Hubbert (1962) used to forecast US oil supply. Hubbert's model is often called a Hubbert curve. The basic idea behind the Hubbert curve is that the supply of oil for a given region will start low, increase, reach a peak and then decline. In this paper I create a simplified Hubbert curve as a sort of "black box" that policy analysts can use to determine oil supply. Analysis of the US lower 48 states shows that its Hubbert curve is not greatly affected by price. Therefore, the Hubbert curve can be thought of as a limit of production. The new "black box" Hubbert curve model will create a non-time dependent Hubbert curve, which will make the curve more flexible for analyzing regional supply issues and even the world supply.

One interesting use for this new Hubbert model will be to analyze world oil supply potential. For example Campbell and Laherrere (1998) give an estimate of world oil supplies using a time dependent Hubbert curve. It may be easier to do a similar analysis using a non-time dependent Hubbert curve and show exactly where the world supply is relative to the Hubbert curve and then debate the issue of the size of the Hubbert curve separately from the point in time that demand will hit the Hubbert curve. Also it is important to be able to use a generic Hubbert curve and analyze how powerful price is for changing that curve.

M. King Hubbert (1962) specifies his logistics model as

$$QP = a \cdot URR \cdot \{1+\exp[a(t-t_o)]\}^{-2} \cdot \exp[-a(t-t_o)]$$

Where QP is the current rate of production; URR is ultimately recoverable reserves; t_o is the year of peak production; and a is a parameter. Hubbert assumes the cause for oil production to increase in the beginning stages of exploration and production is due simply to increasing demand. As demand increases, so does production. The cause for the decline in production is due to scarcity. Once scarcity sets in, oil production reaches a peak and then declines.

An alternative approach to explaining Hubbert's small-large-small sequence of discovery and production is suggested by Georgescu-Roegen's (1972) concept of bonanza where when a natural resource is first exploited, it appears to be a small amount but then slowly is revealed to be much larger than expected. Norgaard (1990) builds on this concept to create an analysis called the "Mayflower Problem." What Norgaard shows is that in the early stages of resource exploration, the resource exploiters do not know where all the best and cheapest resources are. Consequently, they are forced to produce only small amounts of any given resource until more information is revealed to allow them to produce larger amounts of the resource. The information that Norgaard is talking about is not information about new technology (technological diffusion) or information about geophysical concepts (understanding of geophysics), but simply the information about the possible locations of resources similar to a game of hide and go seek. He is assuming in the analysis that technology, technological diffusion, and knowledge of geophysics are all static and constant during the exploration process. This allows the separation of the effects of technology and geophysical understanding from the effects of pure information gathering. Eventually, scarcity causes costs to increase and production to decrease in later periods.

In this paper, cumulative discovery is used as an information and depletion indicator. Technology effects can be added to information effects to further enhance a model. This paper will use the Reynolds (1999) and Al-Jarri and Startzman (1999) model of a non-time dependent Hubbert curve based on cumulative discovery and cumulative production. Price and other variables are added to determine how powerful price is for changing the Hubbert supply model both for discovery and production.

A NEW USE FOR THE HUBBERT CURVE

Hubbert's first model of oil supply, the Hubbert curve, was criticized by Ryan (1965) among others, so Hubbert (1967) created a different analysis using yield per effort (YPE). Cleveland and Kaufman (1991), Cleveland and Kaufman (1997), and Kaufman, Gruen and Montesi (1994) extended Hubbert's YPE analysis and created a model that links current YPE to commutative effort, price and the current rate of effort. The purpose of these models is to determine more explicitly the cause and effect of oil supply. They link the price of oil to exploration activity and then look at the relationship between exploration and actual discovery. While this is an important question, an alternative is to simply look at the relationship between price and the final output supplied. That way a policy maker, without training in the oil industry can then use a black box quadratic Hubbert curve and more easily make decisions relevant to national and regional needs.

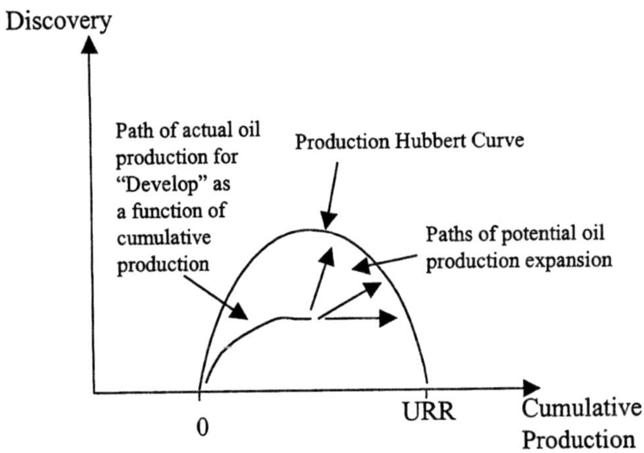

Figure 2.1. The Production Hubbert Curve With Low Oil Production The quadratic production Hubbert curve is a supply limit that the country "Develop" can expand up to.

To see the flexibility of a quadratic Hubbert curve, assume there is an oil producing country called "Develop." For political reasons Develop has kept its oil production regulated at very low levels. However, now Develop would like to open up its oil production regions to competitive market forces and it is interested in determining what its supply potential really is. Figure 2.1 shows Develop's hypothetical Hubbert curve as a function of cumulative production as well as its past oil production. Notice that actual oil production is well below Develop's Hubbert potential. Using the quadratic Hubbert curve, Develop can tell that it currently has the potential to double its oil production if it so desires. Naturally the quadratic Hubbert curve does not show the dates for when production can reach the Hubbert curve or for when production reaches any given point of time. However, dates can be added in as side notes. The relevant point is that now Develop knows where it is relative to its Hubbert supply potential and can better plan where it wants to be. This analysis can be further used to determine where the world Hubbert curve supply is. If the worldwide production of oil is well below the quadratic Hubbert supply similar to Figure 2.1, then we may have potentially more oil than Campbell and Leherrer (1998) suggest. This is certainly what the US EIA (2000) believes.

However, the question still remains, how much of a supply limit is the Hubbert curve? Is price a powerful force for increasing supply above the Hubbert curve trend or is it weak? If price were powerful then we would expect that the elasticity of supply above or below the Hubbert curve with respect to price should be very elastic. On the other hand, if price is not very powerful, then the supply elasticity of oil production above the Hubbert curve with respect to price will be very inelastic. I test the US lower 48 because it is one of the few mature oil regions with a strong competitive market. It is in fact one of the most well adjudicated open markets in the world, and it has quick adaptation to new technology. Any other oil region could easily have poorer characteristics than the US lower 48, but not many could be better. The results here show that two single variables, cumulative production and cumulative production squared, explain 99% of US oil production for almost 100 years, while price is a very inelastic determinant of production. Similar but less striking results can be found for discovery.

OIL EXPLORATION MODEL

Reynolds (1999) subsumes the time variable of Hubbert's original time dependent logistics curves. Both Reynolds (1999) and Al-Jarri and Startzman (1999) show a non-time dependent Hubbert model as

$$L = a \cdot CQD - a \cdot CQD^2/URR$$

Where L = The Hubbert trend; CQD = cumulative oil discovery to date for all previous periods for the entire region; a = a size parameter, (this determines the height and width of the Hubbert curve); URR = ultimately recoverable reserves, a function of technology and regional characteristics.

The Hubbert trend L drives the mineral supply model. However since technology is always changing, URR and production may change. I will use dummy variables to capture this change.

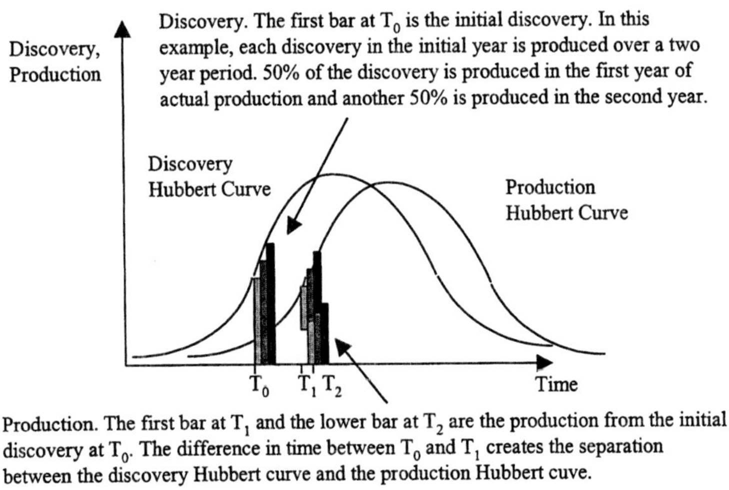

Figure 2.2. The Discovery and Production Hubbert Curves How the discovery Hubbert curve causes the production Hubbert curve to follow the same pattern as discovery

The production of oil follows directly from the discovery of oil. Once oil is discovered, if the pattern of production is the same for all discoveries or at least close to the same, then production follows the same trend as discovery. In Figure 2.2, production follows a simple two period extraction after discovery. Each year a discovery is made, the oil is produced in a subsequent two period schedule.

Each of the discoveries causes the same production schedule making the production trend similar to discovery but offset by a certain amount of time as Hubbert suggested. In this example, production follows a two-year schedule but any other schedule will cause the same effect whether it is a decline curve, constant production, or a multi year schedule. The discovery trend is then the major cause for how the production trend works. Therefore, it is important to determine the effect of price on discovery as well as production.

EMPIRICAL RESULTS

I test the effect of price on US oil discovery and production over time. Yearly production, discovery, and wellhead prices for the US lower 48 states are from the American Petroleum Institute (API), and Energy Information Administration (EIA) data. Data starting in 1900 is used for greater robustness. Discovery data from API is only used until 1978 and after 1978 from the EIA. EIA used different reserve measuring techniques so that the data switch from API to EIA is not congruous. A dummy variable is used to test if there is any affect during the switch where the switch is made before the oil price shock of 1979/1980. The US EIA (1984) created a linkage for the API and EIA data so that one congruous set of data can be used through the data change regime. I do not use that linkage here because the dummy variable will be able to be a proxy for technology and regulation changes as well as the change in data gathering techniques. What we are left with is a pure effect of the ability of price to affect the API data series and the EIA data series independently. The effect of the 1973 price shock affects on the API data and the effect of the 1979/1980 price shock affects only the EIA data.

In the 1970's, due to regulation, there were two wellhead prices for oil. For the discovery model, an average wellhead price determined by EIA is used. CPI is used to find real prices for oil since the incentive to discover and produce oil is a function of profit that can be measured in what the oil explorer can buy directly with his profits. PPI is more a measure of what other firms have to pay for oil, which is a part of their costs. We want to measure oil discovery and production as a profit, which is a direct income to stockholders or wildcat drillers. However, all tests were also made using PPI with no difference in the results.

The hypothesis we wish to test is twofold. First, cumulative discovery or production is tested to see whether they are significant determinants for the current level of discovery or production. Second, the price per barrel of oil as well as Texas Railroad Commission policies, see Kaufmann (1991), are tested to see if they affect the level of discovery or production. One problem with the data is the two tier pricing regulations of the 70's for US domestically produced oil. I use the average wellhead price from EIA in our data. Certainly price regulations of oil must have caused some reduction in supplies, but they cannot have caused the entire supply decline since producers could produce as much oil as they wanted to with the higher price if it was new oil. There was plenty of incentive to expand oil production if more oil was there to produce.

The price or value of oil in the ground is the rent. Price is a good proxy for rent since Adelman (1992) suggests that rent is roughly one third of price and whenever prices have risen or fallen, rent has also risen or fallen usually proportionately and closely in tandem with price. I also consider the possibility that there are lags in the effect of price. For example, the price in one year may affect new wildcat drilling that eventually affects proven reserves in subsequent years due to development drilling. P_0 is current year price, P_{12} is an average oil price for one and two years past, and P_{35} is an average oil price for three to five years past. I test for structural change in the data in 1979 when data collection changed from API to EIA and when prices and market regulations also changed. Adding these changes the model for oil discovery as follows.

$$QD = \alpha_1 + \alpha_2 \cdot P_0 + \alpha_3 \cdot P_{12} + \alpha_4 \cdot P_{35} + \alpha_5(CQD) + \alpha_6(CQD^{(1+\mu)})$$
$$+ \alpha_7 \cdot DUM_{79} + \alpha_8 \cdot DUM_{79} \cdot CQD + \varepsilon$$

Where $\alpha_1, \alpha_2, \alpha_3, \alpha_4, \alpha_5, \alpha_6, \alpha_7, \alpha_8$ = empirical parameters; P_0 = Current Price of oil; P_{12} = The average price of oil one and two years past; P_{35} = The average price of oil three to five years past; DUM_{79} = The dummy variable constant for the structural change in 1979; $DUM_{79} \cdot CQD$ = The dummy variable slope for the structural change in 1979.

Oil production will also follow a similar pattern as discovery, since discovery determines production. The time pattern of production will follow exactly the same path as the time pattern of discovery as long as production follows on the heels of discovery. Therefore the cumulative production path of production will follow the same path as the cumulative discovery path of discovery.

Assume there are no Hotelling (1931) expectations where producers slow down production in anticipation of higher prices, Watkins (1992). Also lagged price variables P_{12} and P_{35} are added to show the effect of previous years discovery on production. The variable TRC is added to take into consideration regulations in Texas that cut down on production capabilities in the US. In addition, dummy variables are added to catch any market structural changes in 1973, 1980 and 1990. In 1973 there was the first oil shock, in 1980 there was the second oil shock, and 1990 was the lead up to the Gulf War and another price shock. The significant price changes in those years induced structural changes in the market such as new technologies and changes in regulation. The dummy variables catch those structural changes. The equation for production is similar to that of discovery,

$$QP = \beta_1 + \beta_2 \cdot P_0 + \beta_3 \cdot P_{12} + \beta_4 \cdot P_{35} + \beta_5 \cdot TRC + \beta_6 \cdot (CQP) + \beta_7 \cdot (CQP^{(1+\mu)}) + \beta_8 \cdot DUM_{73} + \beta_9 \cdot DUM_{73} \cdot CQP + \beta_{10} \cdot DUM_{80} + \beta_{11} \cdot DUM_{80} \cdot CQP + \beta_{12} \cdot DUM_{90} + \beta_{13} \cdot DUM_{90} \cdot CQP + \varepsilon$$

Where $\beta_1, \beta_2, \beta_3, \beta_4, \beta_5, \beta_6, \beta_7, \beta_8, \beta_9, \beta_{10}, \beta_{11}, \beta_{12}, \beta_{13}$ = empirical parameters; CQP = cumulative production; P_0 = Current Price of oil; P_{12} = The average price of oil one and two years past; P_{345} = The average price of oil three, four and five years past; TRC = Fraction of days per year of Texas production allowed due to the Texas Railroad Commission; DUM_{73} = The dummy variable constant for the structural change in 1973; $DUM_{73} \cdot CQP$ = The dummy variable slope for the structural change in 1973; DUM_{80} = The dummy variable constant for the structural change in 1980; $DUM_{80} \cdot CQP$ = The dummy variable slope for the structural change in 1980; DUM_{90} = The dummy variable constant for the structural change in 1990; $DUM_{90} \cdot CQP$ = The dummy variable slope for the structural change in 1990.

Figure 2.3. US lower 48 Oil Discovery and Real Oil Price as a function of Cumulative Discovery

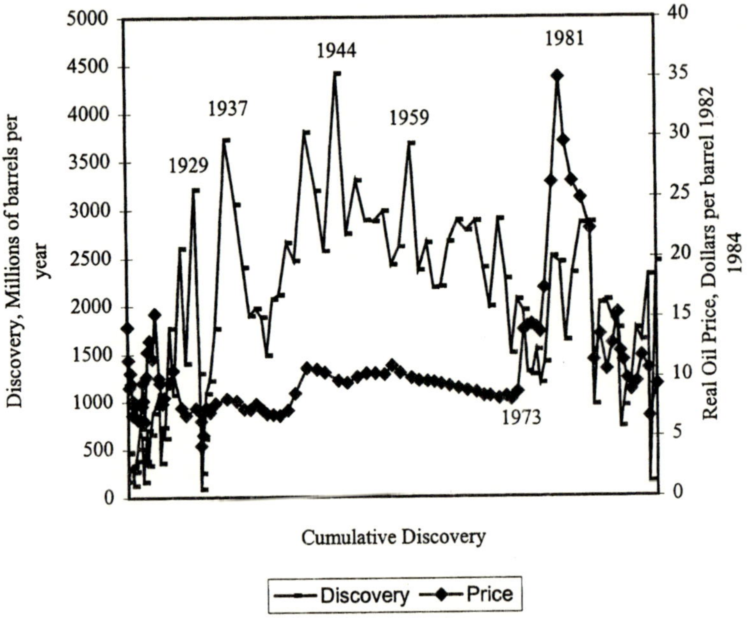

Figure 2.3 shows discovery and real price as a function of cumulative discovery. Table 2.1 shows the OLS estimated results for discovery. I test for auto correlation and find it present. I correct for it with an auto regressive coefficient of order one. We would expect a wide variation in discovery with a somewhat low R^2 due to the possibility of sudden surges in the value of information possibly followed by droughts of valuable information. Yet R^2 is fairly high at 0.7.

Table 2.1. US Lower 48 Oil Discovery
Econometric Analysis of Discovery, 1900 to 1999,
Discovery is defined as increases in proven reserves of Crude Oil
Data from the Energy Information Administration (EIA)
and the American Petroleum Institute (API)

F_{Hetero} (1)	0.112		0.139	
LM F (1)	2.55		1.67	
DW	2.02		2.00	
DF	91		93	
R^2	0.72		0.71	
Augmented Engel-Granger	-5.50* [a]		-5.66* [a]	
		t-statistics		t-statistics
Constant	-1031	-2.94*	-789	-2.84*
CQD	0.08057	10.44*	0.0800	10.52*
CQD^2	-5.16E-07	-8.88*	-5.11E-07	-8.94*
P0	102	3.60*	70.3	3.01*
P12	-54.9	-1.84	----	----
P35	50.3	1.98	----	----
DUM79	-15384	-4.04*	-14161	-3.90*
DUM79* CQD	0.107	4.29*	0.100	4.18*
AR(1) ρ	0.203	1.92	0.184	1.77

(* denotes statistical significance at the 1% level, a denotes statistical information from Davidson and Mackinnon (1993))

The current price is significant for discovery but not lagged prices. However, the price parameter is very low creating an extremely inelastic effect of price on discovery. The dummy variable shows a structural break in 1979. This could have been due to a price change or to the change in data collection. Tests for structural breaks in 1973 or 1990 show no significant change. The model is corrected for auto correlation with an AR1 variable. The reason for the volatility of discovery is due to the fact that information about oil deposits often come in waves. It should be noted that CQD is a quantity where as QD is a rate which makes the two independent from each other in time.

According to Hendry and Juselius (2000) all energy economic data need to be tested for stationarity using tests such as an augmented Dickey-Fuller test, Dickey and Fuller (1979). According to Perron (1989), an augmented Dickey-

Fuller test will show a unit root when there is a structural change in the data. Since there is a structural change in the data, I use an augmented Engle-Granger test, Engle and Granger (1987). The test results show no unit root.

Figure 2.4. US Lower 48 Oil Production and Real Oil Price as a Function of Cumulative Production

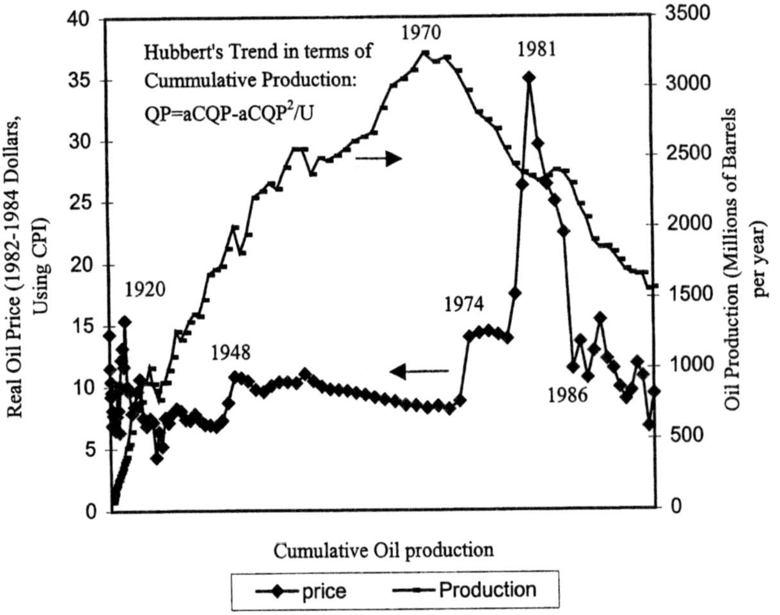

It is especially interesting to look at production of oil. The production of oil in a competitive market would generally begin soon after discovery occurs and require a fairly long period of time to complete the extraction of an incremental increase in discovery. Therefore, production tends to smooth out the variability of discovery. The actual trend of discovery becomes much clearer using the production trend. Figure 2.4 shows production and real price as a function of cumulative production. Table 2.2 shows the estimated results for production. As expected, the R^2 is very high-roughly 99% of production rates are explained by the long-run variables of the model as Hubbert has shown. Dummy variables for

1973, 1980 and 1990 also are significant. We would expect those points in time to cause higher production since they are times when prices changed significantly. However other structural changes occurred then also, such as changes in regulatory policies.

Table 2.2. US Lower 48 Oil Production
Econometric Analysis of Production, 1900 to 1999,
Production is defined as extraction of Crude Oil
Data from the Energy Information Administration (EIA)
and the American Petroleum Institute (API)

F_{Hetero} (1)	0.72		0.139	
LM F (1)	1.48		1.67	
DW	1.82		2.00	
DF	86		88	
R^2	0.997		0.71	
Augmented Engel-Granger	-5.62* [a]		-5.65* [a]	
		t-statistics		t-statistics
Constant	-787	-6.03*	-745	-6.02*
CQP	0.0669	20.95*	0.0670	20.34*
CQP^2	-3.41E-07	-11.04*	-3.43E-07	-10.74*
P0	0.894	0.322	----	----
P12	3.39	1.08	----	----
P35	16.5	4.16*	16.4	4.11*
TRC	655	7.14*	650	7.06*
DUM73	4236	4.79*	3948	4.58*
DUM73* CQD	-0.0428	-4.94*	-0.0399	-4.74*
DUM80	-4612	-3.72*	-4094	-3.73*
DUM80* CQD	0.0386	3.77*	0.0344	3.76*
DUM90	-3943	-2.60**	-3922	-2.58**
DUM90* CQD	0.0281	2.64**	0.0279	2.61**
AR(1) ρ	0.741	9.91*	0.751	10.09*

(* denotes statistical significance at the 1% level, ** denotes statistical significance at the 2.5% level, [a] denotes statistical information from Davidson and Mackinnon (1993))

Noticed that the lagged price is a significant determinant of production, rather than current price. However, price is again a very inelastic determinant of production. For example in year 1970 production was 3241 million barrels per year, average price, P_{35}, was \$8.9 in 1983 dollars and β_4 was 17.5 million barrels per year per dollar, so that the price elasticity of supply at that point in time was 0.05. Simple inspection of Figure 2.4 shows that there was a quadrupling of oil prices in 1973 and yet no quadrupling of oil production any time after that. There was a doubling of oil prices after 1979 and yet there was no doubling of oil production any time after that. In other words supply is extremely inelastic and is so all along the Hubbert trend. I test for auto correlation and find it present. I correct for it with an auto regressive of order one coefficient. I attempted to find other variables that might explain the auto correlation such as the price and production of natural gas, the price of coal as a substitute, and the growth of the economy but no other variables were significant.

I also consider the null hypothesis that the time series has a unit root with a possible nonzero drift against the alternative that the process is trend stationary. The trend though is not time, but the quadratic Hubbert curve of cumulative discovery (production) and cumulative discovery (production) squared. However, since there are structural changes during the trend, those one time breaks in the data can give the appearance of a unit root using an augmented Dickey-Fuller test. Perron (1989) shows that such tests in the presence of structural change will give the appearance of a unit root when no such root exists. To get around this problem, I use an augmented Engle-Granger test, which shows no unit root. Fluctuations are indeed stationary around a deterministic Hubbert trend function.

CONCLUSION

The results show that cumulative discovery explains approximately 75% of the quadratic Hubbert curve for discovery over time for the US lower 48, whereas the price of oil is an inelastic factor. We would expect a less than perfect R squared for the cumulative discovery to discovery relationship because of the volatility of information. Valuable information comes and goes in unpredictable leaps and droughts.

The results regarding oil production are more robust. Since oil production tends to smooth out the variability of discovery, we should expect to find a much closer relationship between oil production and the independent variables, and we do. Roughly 99% of oil production over time is explained by cumulative oil production. As in discoveries, however, the price of oil is a fairly inelastic factor in determining production and price only affects production after a lag of several years. There is evidence that another factor is diverting oil production above the Hubbert trend due to structural breaks in 1973, 1980 and 1990. The reason for this may be due to relaxing of regulations in the oil industry or other changes in the market or technology changes. These structural changes only cause slight variations of Hubbert's original model. Considering the complexity of the oil industry, the changes in technology, and the changes in the economy itself, for almost 100 years, it is amazing to have such a fit.

The quadratic Hubbert curve is found to be an inelastic limit of the supply of oil. Therefore, an underdeveloped country or region that is below the Hubbert curve potential can expect to be able to increase production up to the Hubbert curve with a competitive market structure similar to the US as in Figure 2.1. However, if a country is producing oil at the Hubbert curve it cannot easily increase production higher. The quadratic Hubbert curve then is an inelastic limit to production. Given the analysis of world oil supplies by Campbell and Laherrere (1998), it is important to determine where exactly the world is in relation to the Hubbert curve. This model gives a different way to describe that relationship.

REFERENCES

Adelman, M.A (1992). "Finding and Developing Costs in the United States, 1945 -1986" In John R. Moroney Eds. Advances in the Economics of Energy and Resources Volume 7, JAI Press pp. 11 - 58.

Al-Jarri, Abdulrahman S. and Richard A. Startzman (1999). "US Oil Production and Energy Consumption: A Hubbert Modeling Approach to Forecast Long Term Trends in Various Components of US Energy Consumption with an Emphasis on Domestic Oil Production" in J.R. Moroney ed., Advances in the Economics of Energy and Resources, Volume 11, Fuels for the Future, JAI Press, Stamford, Connecticut, pp. 37 - 58.

Campbell, Colin J., and Jean H. Laherrere, (1998). "The End of Cheap Oil," March Scientific American, pp. 78 - 83.

Cleveland, Cutler J. and Robert K. Kaufmann (1991). " Forecasting Ultimate Oil Resources and its Rate of Production: Incorporating Economic Forces in the Models of M. King Hubbert" The Energy Journal Volume 12, Number 2, pp. 17 - 46.

__________ and Robert K. Kaufmann (1997). "Natural Gas in the US: How Far Can Technology Stretch the Resource Base?" The Energy Journal, Volume 18, Number 2, pp. 89 - 108.

Davidson, Russel, and James G. Mackinnon. (1993). Estimation and Inference in Econometrics, Oxford, Oxford University Press.

Dicky, D.A. and W.A. Fuller. (1979). “Distribution for the estimators for auto regressive time series with a unit root.” Journal of American Statistics Association, Volume 7, pp. 427 - 431.

Engle, R.E. and C.W.J. Granger. (1987). “Cointigration and error cirrection: representation, estimation, and testing.” Econometrica, Volume 55, pp. 251 - 276.

Georgescu-Roegen, Nicholas, (1972). Energy and Economic Myth. in: Nicholas Georgescu-Roegen, Energy and Economic Myths: Institutional and Analytical Economic Essays. Published 1976, Pergaman Press, New York, 3 - 36.

Hendry, David F. and Katrina Juselius (2000). “Explaining Cointigration Analysis: Part I.” The Energy Journal, Volume 21, Number 1, pp. 1 – 42.

Hotelling, Harold (1931). "The Economics of Exhaustible Resources," Journal of Political Economy, Volume 39, Number 2, April, 1931. pp. 137 - 175.

Hubbert, M.K. (1962). Energy Resources, A Report to the Committee on Natural Resources: National Academy of Sciences, National Research Council, Publication 1000-D, Washington, D.C., pp. 54, 61, 67.

__________ (1967)."Degree of Advancement of Petroleum Exploration In The United States," American Association of Petroleum Geologists Bulletin, Volume 51, Number 11, pp. 2207 - 2227.

Kaufmann, Robert K. (1991). "Oil Production in the lower 48 States: Reconciling Curve Fitting and Econometric Models" Resources and Energy, Volume 13, pp. 111 - 127.

__________, William Gruen, and Ronold Montesi. (1994). "Drilling Rates and Expected Oil Prices: The Own Price Elasticity of US Oil Supplies " Energy Sources, Volume 16, pp. 39 - 58.

Norgaard, R.B. (1990). "Economic Indicators of Resource Scarcity: A Critical Essay," Journal of Environmental Economics and Management, Volume 19, Number 1, July, pp. 19 - 25.

Perron, Pierre. (1989). "The Great Crash, The Oil Price Shock, and The Unit Root Hypothesis," Econometrica, Volume 57, Number 6, November, pp. 1361 - 1401.

Reynolds, Douglas B. (1999). "The Mineral Economy: How Prices and Costs Can Falsely Signal Decreasing Scarcity" *Ecological Economics*, Volume 31, Number 1, pp. 155-166.

Ryan, J.M. (1965). "National Academy of Sciences Report on Energy Resources: Discussion of Limitations of Logistic Projections," Bulletin of the American Association of Petroleum Geologists, Volume 49, Number 10 pp. 1713 - 1720.

US Energy Information Agency, (2000). Long Term World Oil Supply, EIA 7/28/2000, found on web page: http://www.eia.doe.gov/pub/oil_gas/petroleum/presentations/2000/long_term_supply/index.htm.

______________ (1984). Two Approaches to the Linkage of US Oil and Gas Reserve Estimates, DOE/EIA-0452, July 1984, Washington D.C..

Watkins, C. J. (1992). "The Hotelling Principle: Autobahn or Cul de Sac?" The Energy journal, Volume 13, Number 1, pp. 1 - 24.

Chapter 3: Oil Exploration Game With Incomplete Information: An Experimental Study*

ABSTRACT

This paper presents a Monte Carlo simulation, using economics students, that shows the characteristics of oil exploration. Oil discovery in the US started low, increased and then declined again. One explanation for this small-large-small sequence of discovery is the role of information. Initial information scarcity causes high exploration costs and low discovery. Eventually information increases pushing exploration costs down and discovery up. Finally, scarcity of resources pushes costs back up again and discovery down. The experimental economics simulation here, shows this to be the case.

INTRODUCTION

Oil discovery in the US lower 48 states has a characteristic small-large-small sequence of discovery and production first postulated by M. King Hubbert (1962) and similar to other oil and mineral exploration regions, such as oil exploration and production in Russia and in various states within the US. The most widely used explanation for this small-large-small sequence of discovery

---

and production is a U shaped cost model. The U shaped cost model is where costs of resource extraction start high, then decline and then eventually increase over time causing oil production to increase at first and then decrease inversely with costs. Slade (1982) and Chapman and Roberts (1983) theorize that technology alone causes costs to decline in the early phases of resource extraction and that lower quality resources and scarcity cause costs and prices to go up again. Cleveland (1991) suggests that technological advancement and the discovery of lower cost oil fields is the main reason costs decline in the initial periods of exploration and that low quality resources is the main cause for increasing costs in later periods. Cleveland though does not explain why it is that low cost oil fields are not found early in the exploration process. Fisher (1981) sees a similar situation but adds an information explanation for price and cost decreases. According to Fisher, new information can cause costs and prices to decline but exactly how information changes over time and how it affects cost trends is not clear. Norgaard (1990) does explain a possible trend with information and the scarcity of information in a model called the "Mayflower Problem". The Mayflower Problem is where the Mayflower Pilgrims landed on the most convenient location possible in the US, even though the land they found was not the most productive farming region in the US. However, the Pilgrims did not know where the best region was and it would have been too costly to find this information.

Reynolds (1999) uses the Mayflower problem to explain an exploration cost model for minerals. Scarcity of information to find, not just low cost oil fields, but any oil fields at all is why costs of exploration start high. The probability of finding oil is low due to a lack of information and this low probability causes high exploration costs. Costs then decline with greater information until scarcity of resources causes costs to increase. Technology effects can be added to information effects to further enhance a model. This paper presents an experimental Monte Carlo model using actual subjects to show how this works. The Monte Carlo model presented here shows the characteristic small-large-small sequence of discovery of oil but with constant technology, costs and prices. The only thing that changes in the model is the explorers' subjective probability of discovery, which drives the sequence of discovery.

The equation for the Hubbert model is:

$$Q = a_0 + a_1 \cdot CQ + a_2 \cdot CQ^2 + e \qquad (1)$$

Where Q = the daily quantity discovered or current rate of discovery; CQ = the cumulative amount of discoveries; a_0, a_1, a_2 = equation parameters.

In a Monte Carlo model by Menard and Sharman (1975), they show that if oil explorers in the US drilled random oil wells throughout the US starting in 1859, then the super giant East Texas oil field would have been discovered at the very least by 1902. East Texas was actually discovered in 1930. Their analysis takes into consideration the given historical rate of exploration and the size of the East Texas field relative to the size of the lower 48 states. Other giant and super giant fields also undoubtedly would have been found earlier by merely drilling exploration oil wells randomly. The problem with this analysis is that the modelers know something that the early oil explorers did not know, namely that the East Texas oil field exists. In other words, the Monte Carlo modelers have perfect information that 22 giant oil fields in the US exist before they run their random search simulation. Therefore, they do not have to pay the cost of being wrong if there are no giant fields. However, the early oil explorers did not have this information and did not know if even one giant oil field existed until the first one was found let alone the East Texas field. After they found the first giant oil field, they still did not know if a second one existed and so on. In order for any Monte Carlo model to reflect this uncertainty, it must be played with real subjects with real costs and uncertainties. In the experiment here, subjects playing explorers are given an incentive to find oil but they also must bare a cost if they do not find oil. In addition, they start their search without any information about what exists. The explorers are therefore motivated to try to use all information available as exploration proceeds in order to maximize their probability of success and keep their costs low. This implies that they will not follow a random search as

Menard and Sharmen theorize. A random search without knowing what if anything exists can lead to low probability outcomes and high costs.

The Monte Carlo simulation here is modeled after roulette where there are black and red slots on the roulette wheel for the ball to fall into. Black represents finding oil. Red represents a dry hole. The probability of getting black on a bet that costs the amount of money for drilling an oil well depends on how many black slots there are as a proportion of all the slots. If you were to play the gamble, the expected results are exactly equal to the probability of getting black. Different countries and different regions would have different probabilities of success. For example compare finding oil in Eastern Texas to that of Maine. The probabilities of finding oil in these two states are quite different, Maine would have a very low probability of success and Eastern Texas would have a relatively high probability of success. In other words, there are different roulette tables with different probabilities, i.e. different proportions of red and black slots. Some roulette tables like some regions and countries have higher probabilities of winning a black slot than others.

Assume then that there is a large room full of roulette tables, and that gamblers will come into the room to play the tables. They will tend to congregate around the roulette tables with the highest proportion of black slots, and play those tables. Based on their expected probability of winnings, they will be more willing to bet on the tables with the higher probability of black since those tables will have a higher expected gain and less losses. Now let us make a change in the game. Assume all the roulette tables are covered over before the gamblers enter the room so that they cannot see the roulette tables. Furthermore, the tables are mixed. In that case, the gamblers do not know which tables have a high proportion of black slots and which have a high proportion of red slots. An additional uncertainty can be added where by the roulette tables in the room have different amounts of black slots. The room could have no black slots at all on any of the tables, or the room could have lots of black slots on most of the tables. There could even be only one black slot on each table. The gamblers don't know.

Now the gamblers have no idea what their chances of winning are or even if there is anything worth betting on or not.

As the gamblers come into the room with the covered roulette tables, they must try to garner information about what the tables are like. Do any tables have black slots? Are there a lot of tables with lots of black slots? No one knows. The gamblers wait to see what other gamblers do and watch for their successes and failures. If a gambler makes a bet and does get a black slot, chances are good that at least that particular roulette table has more black slots. The gamblers race to that table and start betting. However as the bets are made, every time a slot is hit, it is removed and not replaced in the Monte Carlo simulation. There are therefore less and less black slots remaining as betting continues. Eventually the black slots are used up and the gamblers will have to leave old tables and start taking their chances on newer tables.

The gamblers do not randomly play on all the roulette tables, as implied by a pure Menard and Sharman Monte Carlo model, because they don't know if there are any black slots on any of the tables. For all they know, all the tables could have only red. There is no way of knowing a priori what the probability is that any black slots exist. In this situation, the gamblers have in their mind a subjective probability of winning at any given table. However, their subjective probability is different than the actual probability of the table. When the gambler goes to a table where previous gamblers won black, he has information to judge the probability that he also will win based on past experience. That is his own subjective probability. A gamblers subjective probability will be higher at a table with a past history of winning bets than with an unknown table where there is no history. The subjective probability of getting black at a new table may be lower or even much lower than the actual probability of getting black at that table. Therefore there is a difference between a gambler's or a mineral explorer's subjective probability of success and the actual probability of success and it is this difference that needs to be modeled.

A Monte Carlo model of mineral exploration that assumes explorers know what the probabilities of discovery are does not account for the lack of information in early periods. Early American oil explorers, like gamblers facing covered roulette tables of unknown distributions, did not know if there was one giant oil field in the US or two or 22. They did not know if the East Texas oil field existed or not. They only knew the results from explorations already made, they then made subsequent explorations based on that past information, just like gamblers will make their bets based on previous bets. If they eventually do try gambling at all the tables, they will eventually use up the black slots due to scarcity. This is the Mayflower problem. At first there is very little information and few bets are made, i.e. explorations for minerals. Then as more information is found, more betting occurs until eventually scarcity brings to an end any chance of finding new resources.

Table 3.1 Oil Exploration Region and Site Characteristics

Region:	Number of Oil Exploration Sites in Each Region:	Per Cent of Sites that are Black (contain oil) from Lowest to Highest:	Number of Black Sites in Each Region, i.e. Sites with Oil:	Number of Red Sites in Each Region, i.e. Dry Hole Sites with no Oil:
Region 1	20	10%	2	18
Region 2	20	10%	2	18
Region 3	20	15%	3	17
Region 4	20	20%	4	16
Region 5	20	25%	5	15
Region 6	20	35%	7	13
Region 7	20	40%	8	12
Region 8	20	50%	10	10
Region 9	20	65%	13	7
Region 10	20	90%	18	2
Totals:	Total Number of Sites: 200	Per Cent of Total Sites with Oil: 36%	Total Number of Black Sites with Oil: 72	Total Number of Red Sites with no Oil:128

THE EXPERIMENTAL SET UP

In order to simulate oil discovery over time, a Monte Carlo model was created with ten oil exploration regions. Each region has twenty sites to make a play, i.e. to drill an exploratory oil well. The probability of finding oil in the ten different regions is given in Table 3.1. The regions are arbitrarily changed around so that they are not in any particular sequence, however the characteristics of each region remain the same. Participants in the simulation are given 150 points to start with. The object of the simulation is to choose sites with oil, i.e. black, in order to try to obtain as many points as possible and have a better chance to win a lottery ticket at the end of the simulation. The lottery ticket is for a chance to win a $100 gift certificate. With 150 points in hand players choose privately up to three sites from any region where they will drill for oil. They do not know which region is which or even whether there are any black sites at all. When a player selects a site, they are told if the site is red, which is a dry hole, or black, which is an oil discovery. If the site is red, they lose 50 points, which is the cost of drilling a well. If the site is black, then the player receives 100 points, which is the net value of finding oil there. I assume all oil sites are exactly equal in size. This assumption could be modified to include some sites with larger points than others to simulate large oil fields as opposed to smaller oil fields. However, larger and smaller oil field simulation can be done by simply having regions with lots of oil sites, and other regions with less oil sites.

The game is played out in several rounds. All players must make their site selections privately in each round. Each player can make up to three selections in each round, or they can pass and make no selections. At the end of the round it is revealed to all the players which sites were chosen by every one and which sites were red and which were black. Therefore, before the next round every one knows where oil was or was not discovered and they can use that information in subsequent rounds to better select black sites. Rounds are continued for one hour or until everyone decides to stop taking a selection. During a round, players that find oil, gain 100 points for each site that has oil, which they can add to their total points in order to have a better chance at winning a lottery ticket for a prize. Players that choose red lose 50 points. Players can also win some and lose some. For example if a player chooses one red and two blacks they gain 150 points. Or if they choose two reds and one black, they gain no points. If a player's first three selections in the first round are all red, then he goes bankrupt for that game and must leave the game with nothing, i.e. with no chance of winning the lottery ticket. Players can also go bankrupt in subsequent rounds if their point total becomes zero.

After each player's turn, the information of his exploration selection is shown to all other players. Therefore, players have information about previous attempts to find oil. Players always have all the information of previous plays and know how many sites have been selected in each region, how many of those previously selected sites are black and how many are red. Players cannot select a site more than once. In other words, once a site has been selected, it cannot be played again, just like when an oil well is drilled, another oil well cannot be drilled in the same place. The Monte Carlo simulation was played one time through by nine different groups of participants. When the game started players did not know what the chances were at all for finding black or how many blacks there were. Before each game there were two practice rounds on a smaller game board to get the players used to playing. From two Political Economy 100 classes there were formed nine groups of between three and seven students. In every game there were at least one or two students that went bankrupt early.

At the end of each game, the students added up their points and the students with the most points had a better chance at winning a lottery ticket using a random drawing program. At the end of the nine games, there was a drawing of the lottery ticket winner from each game for a $100 gift certificate. The more points each student obtained during the game, the better their chance to win a lottery ticket and then get into the final lottery. The students were also given extra credit.

RESULTS

I found that students who were looking for oil almost never drilled purely randomly. They tried to garner information from what was provided by other student's site selections. Students did not tend to find the large region (region 10) right away. Once in a while region 10 got picked early on. However, no matter when region 10 got picked there was hesitancy to aggressively exploit it because so many other regions had shown that even when black was found, red was also nearby. Almost invariably region 10 was never fully picked out until after half way through the game.

A statistical analysis of the results aggregated from all nine games, shows that a Hubbert curve is characteristic of the rate of discovery. Seven out of nine games had a characteristic small-large-small sequence of discovery where the first third of the black sites found had a low rate of discovery, the middle third of the black sites found had a higher rate of discovery, and the last third of black sites found had a lower rate of discovery than the second third. When all nine games were aggregated together by rounds, the result was the same small-large-small sequence. The Hubbert trend was estimated for the aggregated data. However since the games had very low numbers of participants and since bankrupt students left the game and were not replaced, I added a variable for the number of participants. The results of equation 1 aggregated for all the games by round and modified with the number of students were as follows.

$$Q_r = -22.3 + 0.250 \cdot CQ_r + -4.82 \cdot 10^{-4} \cdot CQ_r^2 + 1.00 \cdot S_r + e$$

$$\quad (-2.52) \quad (4.82) \quad (-3.88) \quad (5.04)$$

where N = 25 rounds; DW = 1.44; LM(1) = 1.86; Adjusted R^2 = 0.94; and the numbers in parenthesis are t statistics.

Q is the number of black sites discovered in each round and CQ is the cumulative black sites discovered. The results show that the Hubbert model explains 94% of actual discovery in an experimental setting. There is no evidence of serial correlation using the Breusch-Godfrey test for first order serial correlation. Figure 3.1 shows the final aggregated discovery, probability of discovery and effort as a function of cumulative discovery of black sites.

What we would expect is that in the beginning phases of exploration, explorers would cut down their efforts and be more cautious due to the low probability of discovery. However as Figure 3.1 shows, effort was extremely high in the beginning even though there were practice rounds before the game started. It seems the students either did not understand how risky their play was, or they had a high opportunity cost of time and just wanted to finish the game and get out. In fact quite a few students went bankrupt in the first few rounds.

Figure 3.1 Discovery, Exploration Effort and Success Rates Versus Cummulative Discovery

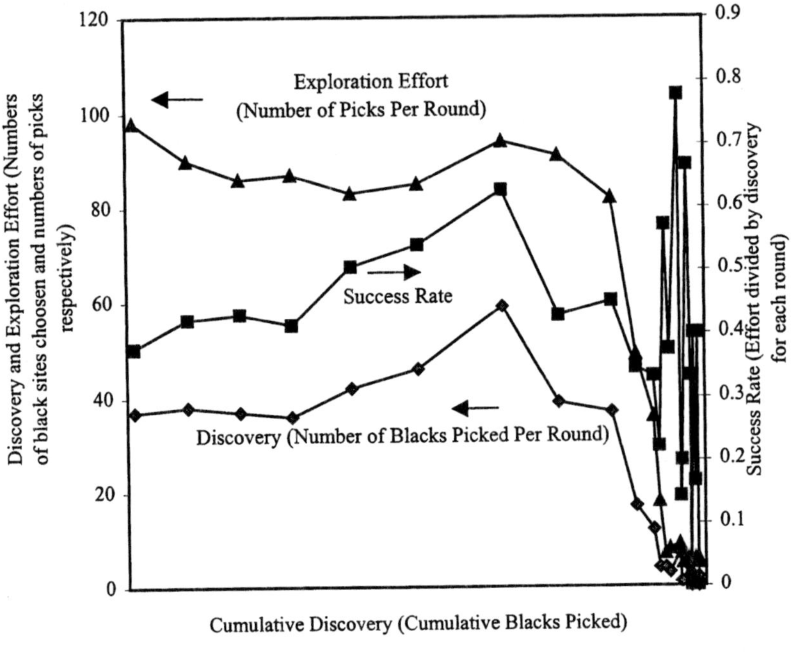

Nevertheless, effort did go down but went up again as probability went up. Theoretically the probability of success for each round should have stayed fairly constant if players rationally tried to maximize their points. However, because of a high amount of effort in the beginning when information was scarce, probabilities of success were low. Probability eventually increased with the information effect. If players were more cautious in the beginning and played the odds better in the middle, what should have happened is that effort would have started low, increased and then decreased and probability would have stayed constant. As it is, probability started low and increased much more than effort. Probability looks to be the biggest factor in creating the discovery small-large-small effect. However, whether probability or effort is the major cause of the discovery movement, the end result is the characteristic Hubbert trend. Either way, it is information that is the driving factor for the Hubbert trend.

In the later rounds, effort, discovery and success rates all went down, although the success rate did become volatile. Usually by the time the last rounds occurred, scarcity of black sites sets in so that it becomes much harder to find more black sites. By the time the games finished, 53% to 75% of the black sites had been found, while only 46% to 64% of all sites had been used. In other words, most of the easiest regions, i.e. regions 10, 9 and 8 were already completely explored by the time the games ended. This meant that there would have been much lower efforts and lower probabilities of success if the games could have continued.

Over all, the low rate of discovery in the beginning round of the exploration game was caused by a scarcity of information. As information increased, it was clear that participants had higher levels of effort and that the probability of success was higher resulting in an increased rate of discovery. Then towards the end of each game, scarcity of resources, i.e. black sites, caused discovery to decline. Efforts declined both due to bankruptcies and due to more cautious play. This shows that the information and depletion effects are key factors in exploration modeling, and that they must be accounted for in any discovery forecast.

CONCLUSION

The result of this experiment shows that information plays a major role in determining discovery. Early rounds of exploration effort were unsuccessful due to a lack of information. Middle rounds were more successful due to better information causing higher success rates. Late rounds were less successful due to depletion. When agents explore for a mineral resource, they naturally gather more information as they go which helps their efforts causing lower cost exploration and greater discovery. However when depletion sets in, discovery declines. The information and depletion effects create M. King Hubbert's characteristic small-large-small sequence of discovery. Certainly price and technology changes can be added to the model to make it more robust. Nevertheless, if technology and prices are held constant, we still get a Hubbert trend. The characteristic small-large-small sequence occurred in seven out of nine games played. The aggregated games show a statistically significant Hubbert trend.

REFERENCES

Chapman, P.F. and F Roberts, 1983. Metal Resources and Energy, Butterworths, London.

Cleveland, Cutler J. 1991. Physical and Economic Aspects of Resource Quality, The Cost of Oil Supply in the lower 48 United States, 1936 – 1988. Resources and Energy, 13: 163 - 188.

Fisher, A. 1981. Resource and Environmental Economics. Cambridge University Press New York.

Hotelling, Harold, 1931. The Economics of Exhaustible Resources. Journal of Political Economy. 39(2): 137 - 175.

Hubbert, M.K., 1962. Energy Resources, A Report to the Committee on Natural Resources: National Academy of Sciences, National Research Council, Publication 1000-D. Washington, D.C., 54, 61, 67.

Menard, H.W. and George Sharman. 1975. Scientific uses of Random Drilling Models. Science. 190(42): 337 - 343.

Norgaard, R.B., 1990. Economic Indicators of Resource Scarcity: A Critical Essay. Journal of Environmental Economics and Management. 19(1): 19 - 25.

Reynolds, Douglas B.1999. The Mineral Economy: How Prices and Costs Can Falsely Signal Decreasing Scarcity. Ecological Economics, 31(1): 155-166.

Slade, M. 1982. Trends in Natural Resource Commodities Prices: An analysis of the Time Domain. Journal of Environmental Economics and Management, 9:122.

Chapter 4: Modeling OPEC Behavior: Theories of Risk Aversion for Oil Producer Decisions*

ABSTRACT

Theories of OPEC such as price leadership, cartel, or game theoretic models suggest an incentive for OPEC members to expand their production capacity well above current levels in order to maximize revenues. Yet individual OPEC members consistently explore for and develop oil fields at a level well below their potential. The cause of low oil exploration and development efforts among OPEC members, and even some non-OPEC members, may have to do with risk aversion. This paper describes an alternative theory for OPEC behavior based on risk aversion using a two-piece Von Neumann-Morgenstern utility function similar to Fishburn and Kochenberger (1979), and Friedman and Savage (1948). The model shows possible low oil production behavior.

INTRODUCTION

In 1973 when oil prices increased dramatically, experts first labeled OPEC as a cartel. Pindyck (1978) theoretically showed how such a cartel could work. Since then Loderer (1985) and others have shown that OPEC did not act as a cartel in the 1970's while admitting there was some evidence of collusion after 1980. Other theories of OPEC behavior were also postulated. Johany (1980) proposed a low discount theory, Ezzati (1976), Cremer-Isfahani (1991) and Teece (1982) proposed target revenue models, Moran (1982) and Adelman (1993) among others proposed political explanations for the cause of the 1970's and 1980's oil price shocks and Al-Sultan (1995) proposed several types of collusive behavior. Griffin (1985) has rejected the low discount and target revenue model hypotheses, although he does show partial sharing evidence. However, Danielsen (1998) shows some cooperation. Gulen (1996) also shows that there is evidence for collusion among OPEC members after 1980. Dahl and Yucel (1996) also show that OPEC countries are non-competitive.

However, the problem with the theory of collusion is that there is always incentive to cheat, but more than that, there is incentive for each country to increase its production capacity in order to strengthen its hand during negotiations. An oil producer who can show a capacity of 10 million barrels a day (mbd) will have a bigger negotiated quota than a producer who can only show a capacity of 1 mbd. If an oil producer can easily expand his capacity from 1 mbd to 10 mbd, but does not, then he may be relegated to a quota of 0.5 mbd when he could have gotten a quota of 5 mbd. Depending on the demand elasticity, the producer could easily triple or quadruple his profits. In addition, Adelman (1986a) shows convincingly that in fact much of the world's oil, albeit OPEC oil, is still low-cost oil in terms of exploration and development. Therefore, most OPEC members could expand their production, but they have not. Even if OPEC is collusive, there is still incentive for individual members to increase their production capacity. A pure cartel or market-sharing model cannot explain why individual OPEC members decide not to expand their production capacity in order to fight for higher market shares within the organization.

One reason that OPEC and some non-OPEC oil producers do not increase their oil production capacity may have to do with risk aversion to exploration and development. Ramsey (1980) shows how exploration and development of oil resources is risky. Friedman and Savage (1948) show that people prefer to accept a certain lower income then to have a higher income with a possible big loss, which is why people buy insurance. This suggests that people are risk averse to losses. However, when people engage in gambling they exhibit risk-loving behavior. Therefore, to a degree, people are risk averse to losses and risk lovers to gains. Markowitz (1952) modified this idea and proposed that people's risk averse or risk loving utility is always relative to current income. In other words, as economic agents' incomes increase, they are still risk averse to losses of their new higher income and they are still risk lovers of increasing their income even more. It is the relative percent gain or loss that people count, not the absolute gain or loss. Fishburn and Kochenberger (1979) show a steeper utility for losses than for gains of uncertain outcomes. Fiegenbaum (1990) shows that firms are risk averse when they are above their targets and risk lovers when they are below their target. This suggests that within a certain range, economic agents are risk lovers to losses and risk averse to gains. This paper tries to incorporate Fiegenbaum's local range of behavior within a larger range of behavior that follows Friedman and Savage.

An alternative model for individual OPEC member behavior and some non-OPEC member behavior can be built incorporating risk averse decision-making. I develop here a model using the Friedman and Savage approach modified by Fishburn and Kochenberger and Fiegenbaum. The model explains how risk aversion causes many OPEC countries and some non-OPEC countries to not increase their production capacity. The analysis suggests that an individual oil producing country can often times gain a higher utility by doing nothing then by searching for or developing oil fields even when the expected net present value (NPV) for such actions is positive.

OPEC DECISION MAKING

Is OPEC a cartel? Griffin states that there is partially effective output coordination, while Dahl and Yucel say that OPEC countries are non-competitive. Therefore, there does seem to be collusion, but that brings up one problem. Why do not OPEC members cheat and try to radically increase each of their shares in production? More to the point, why do OPEC members consistently decide to keep their oil production low? Notice that in official oil production statistics, see Oil and Gas Journal (1970 to 1998), most OPEC members, especially the larger ones, tend to have very high reserve-production ratios, where production can last 50 to 300 years at current reserves for various years. See table 1. By contrast the US and many other free market oil producers have a reserve-production ratio of about only 10 years. Even Norway, which started its oil production in earnest in the 1970's, started with a very high production ratio, but finally reduced it to about 10 years. Even more interesting, some non-OPEC members with country monopolies such as Mexico have very high reserve-to-production ratios. Therefore, there must be a more universal reason behind why OPEC members and some non-OPEC members have such high reserve-production ratios and conversely why they do not increase their production capacity. Adclman (1986b) shows that the reason there are high oil reserves to production ratios is clearly due to a lack of effort into exploration and development by OPEC oil producing countries. Therefore there must be a reason for low production from OPEC countries.

Several hypotheses have been tested to determine OPEC behavior. There are two distinct eras where OPEC behaved differently: the first from 1973 to about 1981 and the second after 1981. Loderer (1985) tests whether official OPEC decisions have any affect on spot market prices. He concludes, "We are unable to find evidence of such an influence (of OPEC decisions on price) for the years 1974-1980 while for the years 1981-1983 the evidence is much more favorable." Gulen (1996) further says, "the 1982-1993 period...is the only period during which the organization has been able to affect prices." Therefore, in the 1973-1982 period OPEC did not affect prices.

If OPEC did not affect oil prices from 1973 to about 1981 then how could OPEC be a cartel? Adelman (1993) suggests that during that time, Saudi Arabia was a price leader and that other countries had political tendencies, which caused lower production and thus the high prices. However this does not explain the reason for countries to sustain high reserve-production ratios when clearly they could have increased revenues dramatically by increasing production. When there is a high reserve-production ratio, then by definition, there is potential to increase production. At any rate there were no official quotas for any OPEC members. All OPEC members were free to produce as much oil as they wanted in that first era. Only in 1981 did OPEC officially begin to coordinate production reductions. The period after 1981 shows that OPEC did affect prices, but not by very much. In fact ironically, it is after 1981 when OPEC had power to affect prices that prices fell dramatically. Therefore, OPEC's power to sustain high prices was very weak.

To understand OPEC, I propose three main models. The first model is a price leadership model. In this model a price leader like Saudi Arabia would have the power to affect oil prices by being the swing producer and would act independently of the rest of OPEC. See Griffen and Neilson (1994). The second model is the cartel model. In this model all OPEC members coordinate their production forming a cartel. All producers would reduce production in order to increase prices and profits, similar to a monopoly. A third model is a game theoretic model. In this model the producers would make agreements to cut production in order to raise prices, but would eventually cheat in order to maximize their own country's revenues. Various strategies would be employed by the members to gain an edge over each other and maximize revenues.

Intrinsic to each of these models there is incentive for most OPEC countries to increase their production capacity. For example in the first model, if Saudi Arabia is a pure price leader as for example a price leading monopolist, then all the other oil producers are free to increase production with no retaliation.

All OPEC members except the Saudi's should increase their production as much as possible to gain the most profit. Yet the evidence shows that almost all OPEC members have had very high reserve-production ratios and have not increased production substantially. This does not make sense in a model of pure price leadership.

In the second model where OPEC is a pure cartel, all members must agree to quotas and not deviate from them. However, from 1973 to 1981, there were no formal quotas made by OPEC for its members. After 1981 quota agreements continually broke down. Griffin and Xiang (1997) show that there is incentive to cheat on such agreements and that indeed members did cheat. Moreover, it is in each member's best interest to increase their production capacity in order to gain a higher share of the profits. There is a huge motivation to push for an ever-higher share by expanding each countries production capacity. Yet again the evidence shows that reserve ratios have often been on the order of 100 years for the biggest OPEC members. There is therefore a lot of potential for those members to increase their production capacity, compared to market economies, in order to obtain a higher share of revenues within the cartel. The incentive structure within a cartel is clear. The members with the highest production capacity are given the highest quotas and so earn the most revenue and profits. Smaller producers earn less. Therefore, there must be some other reason for why members do not increase their production capacity and do not increase it substantially.

The third model is a game theoretic approach where members make agreements on oil production quotas, and then cheat. Saudi Arabia being the biggest producer enforces these agreements, by punishing other OPEC members through over production, which pushes prices down. This is a tit-for-tat approach suggested by Griffin and Nielsen (1994). The problem with this model is that it does not address the issue that producers have an incentive to increase their investment in greater production capacity in order to be able to obtain better quota share agreements. Each producer wants to show that they have a large oil production capacity in order to negotiate a better quota. When producers follow a game theoretic approach where cheating is rampant the expectation is for the cheaters not only to increase production above the agreements but also to increase their capacity to produce above their current capacity. Yet the evidence shows that OPEC members keep their oil production capacity low compared to their reserves.

The lack of strong coordination among OPEC members and the continually low production levels relative to reserves suggests that there is another reason for OPEC member actions. Interestingly some non-OPEC members such as Mexico also exhibit behaviors similar to OPEC members where they have high reserve-production ratios. An alternative model for OPEC members and some non-OPEC members oil production investment decisions would be a risk aversion model along the lines of Friedman and Savage. I explore models of risk aversion for oil producers for small and large firms and country monopolies.

SMALL FIRM UTILITY MODEL

EXPLORATION

A small firm looking for oil is similar to an individual who buys a lottery ticket. Even if the expected net present value (NPV) of buying a lottery ticket is negative causing a loss in income, an individual will often still purchase a lottery ticket. The utility of buying a lottery ticket is greater than not buying it, due to the possibility of great riches. In this same way, the utility of a small firm exploring for oil in a risky area will nevertheless be larger than the utility of not exploring due to the chance of a big discovery that will increase the small firm's income significantly. Figure 4.1 shows the theoretical cardinal utility curve of a small firm. The figure uses cardinal utility instead of ordinal utility in order to show more easily the relationship of income to risk. The same argument can be made with ordinal utility. The figure incorporates Fishburn and Kochenberger's model for utility and risk where near the current income, losses are steeper than gains.

Losses are assumed convex and gains concave similar to Fiegenbaum. However like Friedman and Savage, it is assumed that large gains eventually are also convex due to people's risk loving behavior for large gains. The small firm looking for oil will have a huge gain in income relative to its current income if it finds oil. This is especially true if the firm is a single owner proprietor.

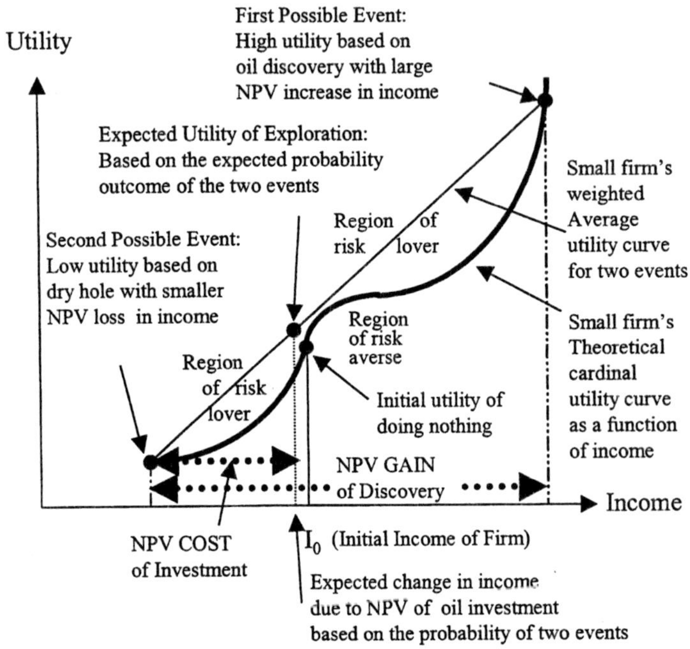

Figure 4.1. Small firm risk utility curve for exploration of a potential new oil field where the field is large relative to the size of the firm

In Figure 4.1 is shown a negative expected NPV of a particular oil exploration project. The cost of the exploration is negative, while the potential income if oil is found is positive. The expected overall gain depends on the subjective probability of success. In this figure is shown a negative expected outcome. It is also possible that the expected NPV of an exploration project is positive in which case there is even more incentive to search. What I want to show here is that even if the expected income is negative, small firms will still search for oil. The utility is high not because of the expected income but because there is a chance of having a large increase in income relative to current earnings. It is a gamble of high stakes where utility is a bigger motivator than the actual expected net present value outcome. However, if there were continually negative expected values for oil exploration, then we would expect that firms would leave the industry due to bankruptcy. However, new firms will constantly enter the industry. In addition, new capital from small investors will normally be injected into ailing companies to keep them going. The small investors, who also have a negative expected value of the investment, will invest their money similar to taking a gamble. They are also interested in the chance of making a large profit.

For example in the late 1920's, C.M. "Dad" Joiner was searching for oil by himself for three years in East Texas before he found the super giant East Texas oil field. He virtually lived in poverty and continually ran out of money during those three years of exploration, but small investors kept buying stocks in his company all hoping for a huge oil discovery and big payoff. This kept Joiner going. This shows that it is possible for companies to actually lose money while exploring for oil due to the high utility of a potential huge discovery relative to their small current income. This has been the case in countless gold, silver or oil rushes, where small firms (mostly individuals) invest every cent they have into traveling to and developing mineral claims in the hope of striking it rich. The expected NPV income of the vast majority of these explorers is negative, yet they search in the hope of striking it rich. In this sense, we expect that small firms will be bigger risk takers in oil exploration. This is due to the relatively large potential payoff of an oil discovery compared to their small current income, and their risk loving utility curve for large payoffs.

DEVELOPMENT

Although at first glance, developing an oil field does not seem particularly risky since we know the oil is there, yet there are in fact many risks and uncertainties associated with the process. First a producer must discover a reservoirs characteristics and how large an area it encompasses. It is possible to drill development oil wells outside of the producing area and lose money. Then the producer firm must decide how many wells to place on a reservoir, how fast they will pump, what type of injection wells if any will be used and other issues. In general, the slower the firm develops the oil field, the more certainty it has that each well or infrastructure investment will maximize profits, although the longer it will take to obtain those profits. There is a trade off between time and risk.

Another risk in developing an oil field is where oil is extracted too quickly, which causes the geologic structure of the reservoir to deteriorate. When oil production proceeds too fast, the total recoverable reserves of the reservoir decline. In other words, higher production today can lead to much lower production in subsequent periods. In order not to lose total recoverable reserves, an oil producer must find out the structure of the reservoir, which requires gaining information about its geological characteristics as the development of the reservoir proceeds. Because of the possibility of adding too many unneeded oil wells, or extracting oil too quickly, there is a risk of unnecessary or detrimental development.

In addition, a further risk for potential oil field development is the possibility that future oil prices will rise faster than expected, causing the oil producer to lose money by not waiting for higher oil prices, or alternatively waiting too long. Risk due to future price expectations also occurs with input prices such as labor and capital. There is even risk associated with changes in laws determining property rights or taxes. All these aspects create at least some risk for developing a given oil field. A good explanation of output price risks can be found in Salahor (1998).

In Figure 4.2, is modeled the potential losses and gains of increasing oil production from a given oil reservoir. What is shown are two mutually exclusive outcome events, a loss outcome and a gain outcome-with a subjective probability for which one will occur. The potential NPV loss is where oil production is too high and therefore causes a significant loss in total oil reserves. The potential NPV gain is simply due to higher production today. The expected gain will be based on the probability of the two events. The subjective probability of the two events for the development of an oil field will more than likely be a very low risk action creating a solid NPV gain. However, even if development were to have a negative expected NPV, a small firm would still be more likely to take the risk of development than not take it.

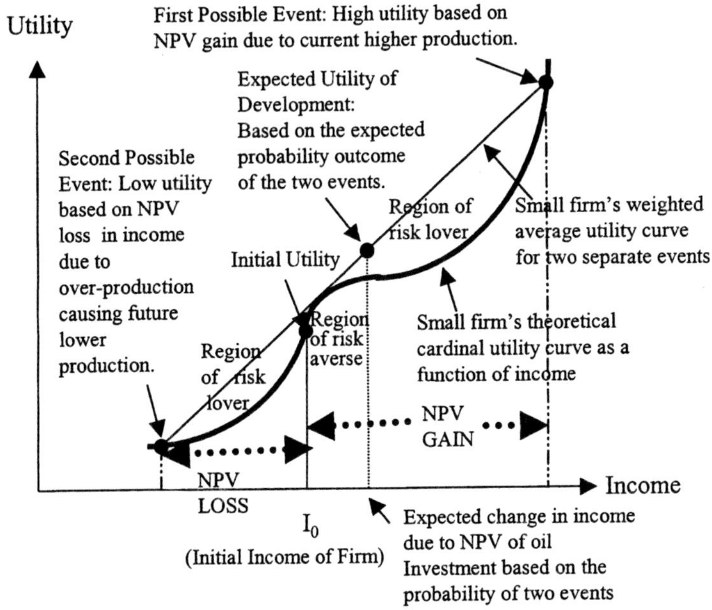

Figure 4.2. Small firm risk utility curve for development of an oil field where the field is large relative to the size of the firm.

LARGE FIRM UTILITY MODEL

A large firm just like a small firm will also exhibit behaviors of an individual and be risk loving to large gains but relatively risk averse to losses. The difference between a large firm and a small firm, though, is that the large firm starts with a much higher initial income. The high initial income of a given firm creates a slightly different behavior for large firms because they require relatively larger gains and losses to change their utility. When a small firm explores for oil, almost any oil field it finds will create a large increase in its income, but for a large firm only large oil fields will significantly increase income. Most large firms will not spend a lot of effort investing into small payoff projects because the utility gain is so small. It is the size of the discovery relative to the initial income of the firm that determines how much utility changes. Small projects are not significant enough to large firms to warrant their attention. However, often times a large firm may simply buy a small firm's oil field mineral rights and infrastructure once all the risks have been taken and the oil field has no uncertainty of income.

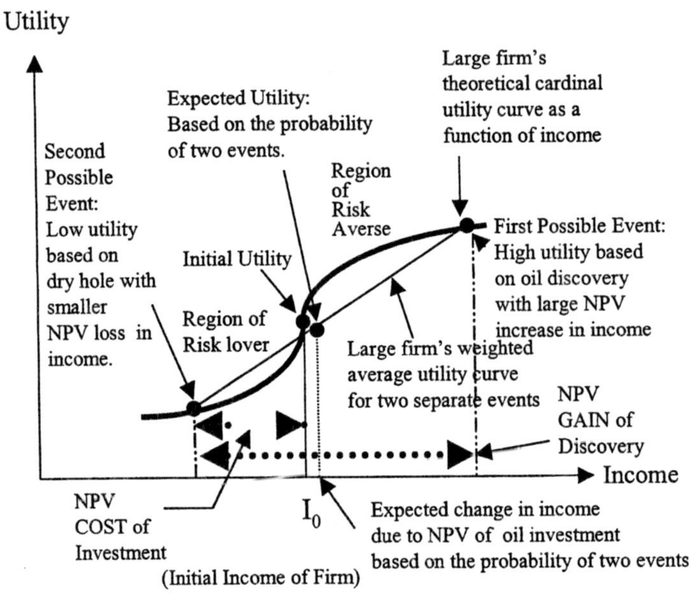

Figure 4.3. Large firm risk utility curve for exploration of a potential new oil field where the field is not large relative to the size of the firm.

Figure 4.3 shows a large firm's utility curve for exploration. Again, as in Fishburn and Kochenberger, losses are steeper than gains and as in Fiegenbaum, losses are convex and gains are concave, although the overall range is mostly concave and risk averse. This curve is flatter than the curve in Figure 4.1 and is only convex for gains because when a large firm makes an oil discovery of a similar size as a small firm's discovery, the relative change in income for a large firm is smaller than for a small firm. This causes the utility of looking for small oil fields by a large firm to be less than doing nothing, even if there is an expected NPV gain. In addition, a large firm often times has higher costs of exploration due to its larger size. It is less efficient at making decisions and moving quickly to take advantage of an opportunity, which increases its fixed costs to start any given project. On the other hand, a large firm has an advantage with economies of scale, where large high cost projects need a lot of initial investment before a payoff can be had. Small firms would not have the capital to do such projects. This may be why large firms tend to concentrate their efforts on larger oil fields and reservoirs and small firms go after smaller oil fields and reservoirs.

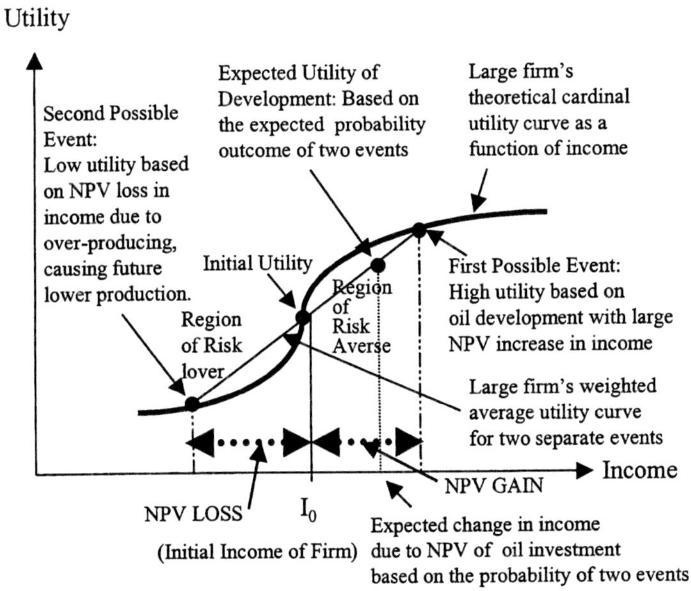

Figure 4.4. Large firm risk utility curve for development of an oil field where the field is not large relative to the size of the firm

Figure 4.4 shows the development of an oil field for a large firm, which is similar to Figure 4.3, but where the subjective probability of success is much higher. Each new investment in developing an oil field will have costs and gains but also some risks. In Figure 4.4, again is shown two mutually exclusive outcomes for a single development decision of an oil field one with a NPV gain, the other with a loss. The loss is due to the possibility of future lost oil production or some other risk factor. The expected NPV gain in income from investing in oil field development is usually pretty high due to the very low risk involved. There is a high probability of success. Still even with the low risk and high expected NPV gain, the utility gain of oil field development for a small oil field will not be particularly high for a large firm. In that case, there may be some hesitation to develop a field even though there is a fairly risk free NPV gain in income. By contrast, smaller firms will explore and develop fields much faster due to the higher expected utility of each investment.

Small firms move faster on projects and take more risks in oil exploration and development due to the higher potential gains in income relative to current income. Larger firms are slower and take less risk. This may explain why a lot of onshore oil exploration and development in the US proceeded fairly quickly, while offshore oil development proceeded more slowly. Offshore oil development has a larger up front cost that only allows large firms to enter the market. The large firms follow a plan of less risk and slower development. The onshore development has a lot of smaller firms that can quickly develop projects and are willing to take a lot more risk.

COUNTRY WIDE MONOPOLY FIRM UTILITY MODEL

In a number of oil producing countries, especially OPEC countries, there is only one Oil Company for that country. Examples of this would be ARAMCO in Saudi Arabia, the National Iranian Oil Company (NIOC) in Iran and even Pemex in Mexico. The host country (HC) either owns a single oil firm for the entire country or the HC may simply contract out all oil exploration and development projects. A third strategy often used is to have joint ventures where an International Oil Company (IOC) takes all the risks of a project but is given a percentage of the profits. Nevertheless, all oil industry decisions within the country are highly controlled by a single entity owned by the state. We can simply call such an entity, the country monopoly. For example, Atsegbua (1999) gives a good history of Nigeria's oil industry regulations where in the 1950's and 1960's IOC's could acquire leases. After 1971, the Nigerian National Petroleum Corporation (NNPC) was formed to control all oil operations. The control by NNPC means IOC's have to negotiate how much profit they will get from successful explorations. The level of reward and the time it takes to go through such negotiations often causes incentives for oil exploration to be too low.

Similar to a large firm, a country monopoly will have a very large initial income. However when comparing a large firm such as Exxon with a relatively small oil producing country such as Algeria, one cannot look at Exxon's total revenue, and compare it to oil revenues of Algeria. Rather one must compare only the profit of Exxon, which is its income, to the entire GDP of Algeria, which is its income. Profit and GDP are the respective relevant relative incomes for these economic agents. For example in 1997, Exxon had a profit of about $7 billion while Algeria had a GDP of about $50 billion so that Algeria is in fact a much bigger economic agent than Exxon even though one might expect Exxon to be a vastly larger economic entity than Algeria. I believe the country monopoly will behave as if its income is the GDP of the country, and therefore it will be similar to a very large firm.

However, a country is quite a political entity unlike a large firm. Even when a dictator runs a country, the country monopoly is subject much more to politics and especially public opinion than is a large firm. See for example Gurney (1996) and Dixit (1996). The politics of a country though do often deal with economic gains or losses to the citizenry or perceived unfair economic treatment usually where one entity gains more economically than another entity or where one entity gains more than the citizenry at large. Therefore characterizing a country as an economic entity is a close approximation to how it works. Figure 4.5 shows the characteristics of a country monopoly for exploration. I believe that since a country monopoly has more political concerns than a private company, it has a very high disutility of losses and a very low utility of gains. In other words, a country's utility for gains is very flat and it's utility for losses very steep.

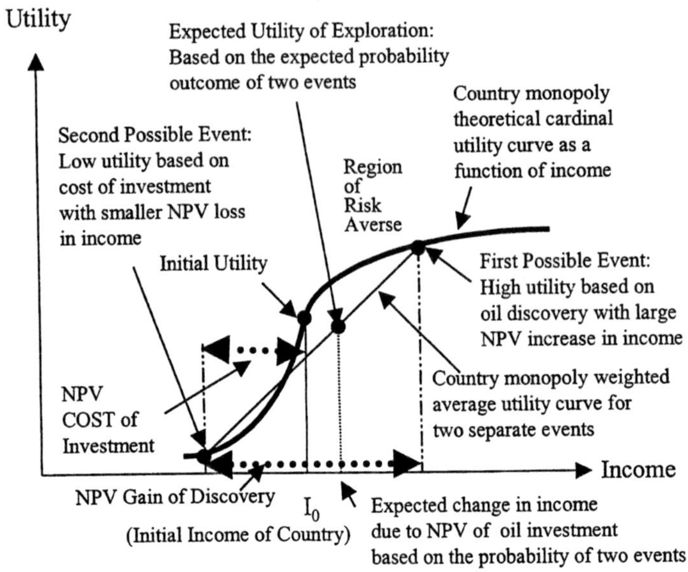

Figure 4.5. Country monopoly risk utility curve for exploration of a potential new oil field where the new field is not large relative to the size of the country GDP

For example, consider gains. When a country has a gain in oil revenue, the gain is quickly dissipated into several agencies of the government, not the least of which are the armed forces. Most citizens of the country do not see any gain. Therefore citizens, and the political body that caters to them, get very little utility from oil revenue gains. The political utility increase is small. Examples of low utility gains for increasing oil production abound in developing countries where local residents near an oil well see very little benefit for themselves of oil development, but then they see oil workers and bureaucrats becoming rich. This creates a lot of resentment among the locals. Often times, local populations have a perception that the oil being produced is being stolen from them by the West, again creating very low utility of oil development. Oil companies are keenly aware of this lack of utility among local populations, and therefore the companies constantly advertise or show how they are hiring local residences, helping local communities near their oil investments and investing in the local economy. Perception is everything in these oil field developments and often it supersedes any NPV gains of exploration and development.

On the other hand, Ramsey (1980) shows how a county's political apparatus is often acutely sensitive to losses, which can be extrapolated to losses caused by a country monopoly losing money. For example, if the government were to lose money on some aspect of oil production, such as drilling an expensive dry well or not waiting for oil prices to go up before producing, then the citizens would become incensed at the waste. A loss in oil revenue might cause the population to get very upset because people feel that the oil industry, even if it is run by their own oil company, is some how stealing from them. Although if the monopoly is inefficiently using its resources to help locals, such as giving away free gasoline to people, that is okay. Also, the oil workers who are rich don't have to suffer as much because of the loss, which is felt as unfair by the general population. In addition, if a country monopoly is found to have made a risky investment and lose some investment money, then all segments of the population will criticize how the government is running its oil operations badly. So in the political utility arena, oil revenue losses are extremely detrimental while oil revenue gains are very diluted. Better to keep a steady income, than to increase it and have a greater risk of future losses.

On top of this, a country monopoly, even if it contracts out to an efficient firm for all its work, will tend to have many layers of bureaucracy. The country monopoly will not have competition to force it to run efficiently and therefore it will tend to have much higher costs of business than other companies. Often times country monopolies or governments themselves spend so much time negotiating agreements with oil firms trying to get the very best deal, and trying not to be taken advantage of, that very little oil is explored for and produced. Also the country monopoly will employ too many workers, too much capital and give its employees extra perks in order to keep as much of the oil revenue for itself as possible before giving away the profits to the government. The incentive is for inefficiency. The net result is that projects run by government companies have much higher costs than are necessary and therefore have a lower expected increase in income for the country. The lower profit further erodes incentive to invest. On top of this, the lower increase will be compared to the entire GDP of the country, which means that it takes an incredibly huge oil field or oil project to induce the country monopoly to do anything at all. Even if the risk of a project is low, the resulting profit may still be too low to have a utility higher than doing nothing. What tends to happen is that a country monopoly will only invest in the lowest risk, highest gain projects possible. Few risks will be taken. This tendency will cause such countries to have very low exploration activity.

An example of how countries do not take risk in exploration is when the country monopoly offers an independent oil company an extremely low reward for finding more oil. See Adelman (1986a). If for example the HC offered only \$0.10 for every new barrel of oil found and it gains a NPV of \$15 per barrel for selling the oil, then it loses virtually nothing by having an IOC explore for the oil. If however, the HC pays \$10 per barrel of new oil, the IOC gets a substantial portion of the profit for finding oil and the HC will lose a large portion of the profits. Such a loss of potential profits for a political entity would create an uproar among the population causing large disutilities. The country monopoly would be very sensitive to this and so want to pay very little for newfound oil. However with a low incentive such as \$0.10 per barrel, very little efforts will be made by an independent oil company to find more oil. Hence only very large oil fields will be searched for and only sporadically.

Adelman (1986) suggests that oil prices are relatively high compared to costs of oil exploration and extraction especially in the Middle East. Furthermore, most OPEC members have official reserve-production ratios well above 50 years, while the US and many other competitive oil producers consistently have oil reserve-production ratios at about 10 years. Therefore, if OPEC did follow a price leadership model or a collusive model, then all its members should still be investing more into their upstream oil production infrastructure in order to be able to claim the right to a higher share of overall OPEC production. Yet they invest very little. Furthermore, non-OPEC oil producers such as Mexico also exhibit low oil industry investment behavior. The reason behind this is the very risk averse nature of the politically minded country monopoly.

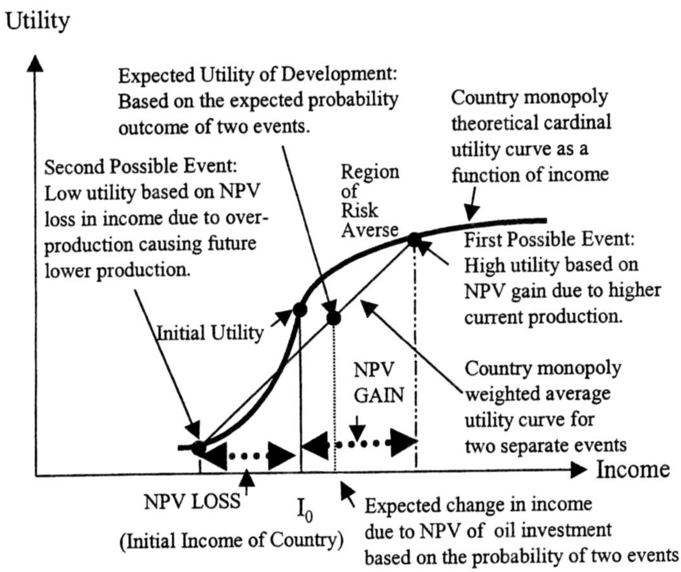

Figure 4.6. Country monopoly risk utility curve for development of an oil field where the field is not large relative to the size of the country GDP

Another possible risk for a country monopoly is during development. In Figure 4.6, we see how this works. If the country monopoly decides to develop an oil field now, then that could mean that future reserves would be lost. In addition, if the price of oil increases in the future, then there could be a loss of NPV revenue due to not waiting for the price increase. On the other hand, there is a current gain of income by increasing production now. The subjective risk between two mutually exclusive events of a loss and a gain determines the expected NPV change in income. It is possible that even if the subjective risk of a project is low and the expected gain in income is fairly high, the country monopoly will be so risk averse that development of oil reserves may still have an overall disutility. This is especially true if the costs of doing a given project are very high as they often are when done by a country monopoly.

In addition, OPEC members not only face a game theoretic repeated prisoner's dilemma, Axelrod (1992), for today's production but also for investment into future production. Investment into oil development itself can create risk of a Saudi reprisal where if any one country decides to invest in oil development, then Saudi Arabia will increase production to push oil prices lower and punish this potential increase in oil production capacity with lower prices. Any OPEC member that would want to increase its production significantly would undoubtedly need 2 or 3 years of intense investment to accomplish a large oil production capacity increase. It would be immediately obvious to Saudi Arabia and the rest of OPEC that such investment would induce cheating on OPEC objectives and the project would force Saudi Arabia to increase production now in order to lower prices and punish that investment. The lower oil prices would then slow down the country monopoly's plans to invest. However, merely the risk of a Saudi reprisal may cause the same decision outcome. Even countries that announce that they are increasing production capacity may not actually be doing it due to risk aversion. It is not what is said but what is actually done that matters.

What may be happening in OPEC is that what looks like coordinated OPEC strategies are really independent but simultaneous risk appraisals. One risk is whether a given OPEC member's own oil investment will cause Saudi Arabia to increase its production and lower prices further. The risk of that happening especially for the larger members and the lack of utility for gains in NPV causes oil producers to become extremely risk averse. In other words, the utility of gains and losses in a repeated game is much different then the dollar values of gains and losses. Countries are naturally politically risk averse and therefore the utility of a loss is much greater than the utility of a gain. This changes the strategy and outcome of any game theoretic analysis of OPEC.

Another risk is where many OPEC members believe oil prices will increase significantly in the near future. They do not want to risk producing a lot of oil in the present under a low price environment and lose future high revenues. In the 1970's and early 1980's many highly respected energy analysts, see for example DOE 1984, predicted ever-higher oil prices. A country monopoly would naturally become very risk averse to increasing oil production today if it expected future high prices. The similar actions of OPEC members look coordinated but may in fact be simultaneous reactions to the same set of risks.

THEORIES OF OPEC REVISITED

Using the risk aversion theory, I can now explain how OPEC works. In the 1973 to 1981 period all OPEC members were free to increase their production as much as they wanted too. Only price ceilings were negotiated. However, Loderer shows that those agreements made no affect on the world spot prices. What may have happened was that Saudi Arabia was a pure monopoly price leader at that time. Saudi Arabià deliberately kept exploration and development low in order to keep oil prices high all though Saudi Arabia itself also undoubtedly had some risk aversion. The rest of OPEC and the rest of the world for that matter could not increase production very much. OPEC members could not increase production because new oil investments would take time to build up. Furthermore, new investments were risky which further inhibited exploration and development. Exploration and development in non-OPEC countries certainly took off during this period, but simply took time to get on line. Some new investments occurred in OPEC countries too but not at nearly the pace and quantities that it occurred in non-OPEC countries.

In the period since 1982, world demand for oil has flattened and even declined. This took away Saudi Arabian monopoly price leadership power, but still allowed it to be the dominant player in an oligopoly. With Saudi Arabia in the lead, OPEC then played a tit-for-tat game theoretic model. OPEC members go through cycles of agreeing to quotas and then cheating. First everyone agrees to a quota. Then members cheat. Then Saudi Arabia punishes the cheating by increasing production and lowering prices. After Saudi Arabia pushes prices lower, OPEC gets back to the bargaining table and tries to cut production again. Finally OPEC makes a new set of agreements that everyone follows. Then members cheat again and the cycle continues.

However, if OPEC members really wanted to cheat, they would change their property rights arrangements so that oil companies big and small could be rewarded for exploration. This would increase investments in oil exploration and development very quickly giving producers more production and revenue visa-a-vie non-cheaters or even slow cheaters. So why do not members cheat more intensely? Each country is risk averse. They each have a natural political risk aversion to making changes that would increase production. They are risk averse to making huge new investments in exploration and development and so their cheating is quite small. The risk aversion model here can therefore explain the dichotomy between the periods before 1982 and after.

CONCLUSION

A lot of theories abound as to what OPEC is and how it acts, but no matter what model is chosen there is incentive for all the producers to increase their oil producing capacity. Yet for years and now even decades, there has been a decidedly lack of investment in creating extra oil production capacity, not just by OPEC members, but even by some non-OPEC members. Even if OPEC follows some sort of Stackelburg equilibrium or a tit for tat strategy among its member, each of the members must have something to fight with. They each need not just excess production but ample excess production in order to be able to strike blows at each other, and substantially increase their own quota and revenue. In addition, cheating is much more profitable if an oil producer creates much more excess capacity. Yet no one creates this capacity.

Many OPEC members have very high reserve-production ratios. Most of these countries should be able to increase production by 2 or 3 times current production if they had competitive oil market environments similar to the US. With so many oil producers, one would expect that the first one to increase their capacity would be the first one to gain the most profit. Yet the producers do not invest. This has continued to occur not just for a few years but for decades. One explanation for this is the result of the natural extreme risk averse nature of any political body. The extreme risk averse nature of oil producing country monopolies inhibits investment into oil exploration and development. In addition, country monopolies face higher risks for investing in production, namely that they will lose future higher revenues if prices rise in the future as some experts predict or that Saudi Arabia will punish their actions with higher oil production and lower oil prices. I believe risk aversion is the main cause of low oil production for oil producing countries with country monopoly firms.

The theory of risk aversion then can explain the seemingly dichotomous behavior of OPEC before and after 1982. Before 1982, Saudi Arabia had a pure monopoly leadership power to increase oil price by keeping its production low, i.e. lower than it could have produced oil. This is why there were no quota agreements but only price agreements. As a price leading monopoly, Saudi Arabia was interested in keeping price high but not too high lest new sources of oil came on line. Unfortunately for Saudi Arabia, new oil sources did come on line and demand did begin to decline. The other members of OPEC kept their production low during this time due to their natural political risk aversion.

After 1982 demand sank low enough and non-OPEC production was high enough that Saudi Arabia lost pure price leadership monopoly power. However, Saudi Arabia could still be a price leader and force the other members to agree on quotas. OPEC began a tit-for-tat game. Even though OPEC members always cheated, they did not cheat too much and cause the price to go down to the competitive floor. The reason they did not cheat too much was because of their natural political risk aversion. They were risk avers to increasing production by larger and larger amounts.

Adelman has said that, "(Oil) supply is being restricted to maintain the price, which can be maintained only so long as the low-cost oil is dammed up. If the dam breaks, so will the price." However, what is the "dam" that has been keeping prices high? It cannot be the political will of OPEC. If OPEC had any kind of political will at all, they would be able to carry out an effective cartel and oil prices would be as high as $100 per barrel today. In fact, it is political fear and intense risk aversion that keeps OPEC members from increasing production more substantially.

While oil prices in 1998 got down to $10 per barrel, about as low as pre-1973 prices in real terms, this is not due so much to competition among producers as a lack of demand. As soon as demand goes up, expect OPEC members and some non-members to continue to keep production low, out of fear of making a wrong move. Oil price shocks in the past, and undoubtedly those in the future, are due less to OPEC coordination then to OPEC risk aversion and high demand. In that case, the "dam" of low cost oil should continue to hold for a very long time to come due to this risk averse behavior. We can also expect that when demand starts increasing again with greater world economic growth, that supply will not increase in tandem, but will stagnate. In the future then, oil will again be high priced.

REFERENCES

Adelman, M.A. (1986a). "Scarcity and World Oil Prices," The Review of Economics and Statistics, August, pp. 387 - 397.

_____________ (1986b). "The Competitive Floor to World Oil Prices," The Energy Journal, 7(4), pp. 9 - 31.

_____________ (1993). The Economics of Petroleum Supply, MIT Press, Cambridge.

Al-Sultan, Abdulrahman M. (1993). "Alternative Models of OPEC Behavior," Journal of Energy and Development, 18(2), pp. 263 - 281.

Atsegbua, Lawrence (1999). "The development and acquisition of oil Licenses and leases in Nigeria," OPEC Review, 23(5), pp. 55 -77.

Axelrod, Robert (1992) "The Problem of Cooperation," in Cowen, Tyler ed. Public Goods and Market Failures: A Critical Examination, London, pp. 237 - 254.

Cremer, Jaczue and Djavad Salehi-Isfahani (1991). Models of The Oil Market, Harwood Academic Publishers.

Dahl, Carol and Mine Yucel (1991). "Testing Alternative Hypothesis of Oil Producing Behavior," The Energy Journal, 12(4), pp. 117-138.

Danielsen, Albert L. (1988). "OPEC Stability: An Empirical Assessment," Energy Economics, 10(3), pp. 174 -184.

Dixit, Avinash K. (1996). The Making of Economic Policy: A Transaction-Cost Politics Perspective, Cambridge, MIT Press.

Ezzati, Ali (1976). "Future OPEC Price and Production Strategies as Affected by its capacity to absorb oil revenue," European Economic Review, August, 8, pp. 107 - 138.

Friedman, Milton and Leonard J. Savage, (1948). "The Utility Analysis of Choices Involving Risk" Journal of Political Economy, August, Volume 56.

Fiegenbaum, Avi (1990). "Prospect Theory and the Risk-Return Association: An Emperical Examination of 85 Industries," Journal of Behavior and Organization, 14, pp. 187-203.

Fishburn, Peter C. and Gary A Kochenberger, (1979). "Two-Piece Von Neumann-Morgenstern Utility Function," Decision Sciences, 10, pp. 503 – 518.

Griffin, James M. (1985). "OPEC Behavior: A Test of Alternative hypotheses," American Economic Review, 75(5), pp. 954 – 963.

__________, and William S. Neilson (1994). "The 1985-86 oil price Collapse and Afterwards: What does Game Theory Add," Economic Inquiry, Volume 32, October, pp. 543 – 561.

__________, and Xiong, Weiwen (1997). "The Incentive to Cheat: An Emperical Analysis of OPEC," Journal of Law and Economics, 40(2), October, pp. 289 – 316.

Gulen, S. Gurcen (1996). "Is OPEC a Cartel? Evidence from Cointegration and Causality Tests," The Energy Journal, 17(2), pp. 43 -57.

Gurney, Judith (1996). Political Economy of Oil Exporting Countries, Volume 4, Oxford University Press, Oxford.

Johany, Ali (1980). The Myth of the OPEC Cartel, N.Y. Wiley and Sons.

Oil and Gas Journal (OGJ). (Various years, 1970 - 1998). "World Wide Look at Reserves and Production," Oil and Gas Journal, December.

Loderer, Claudio (1985). "A Test of the OPEC Cartel Hypothesis: 1974 - 1983," Journal of Finance, 60(3), July, pp. 991 - 1008.

Markowitz, Harry (1952). "The Utility of Wealth," Journal of Political Economy, 60 April, pp. 151 - 158.

Moran, Theodore (1982). "Modeling OPEC Behavior: Economic and Political Alternatives," in Griffen and Teece, eds. OPEC Behavior and World Oil Prices, London, George Allen and Unwin, pp. 94 - 130.

Oil and Gas Journal. "World Wide Petroleum Reserves," Oil and Gas Journal, various years.

Pindyck, Robert S. (1978). "Gains to Producers from the Cartelization of Exhaustible Resources," Review of Economic Statistics, 60(2), pp. 238 - 251.

Ramsey, James B. (1980). Bidding and oil Leases, JAI Press, Greenwich, Connecticut.

Salahar, Gordon. (1998). "Implications of Output Price Risk and Operating Leverage for the Evaluation of Petroleum Development Projects," The Energy Journal, 19(1), pp. 13 - 46.

Teece, David J. (1982). "OPEC Behavior, an Alternative View," in Griffen and Teece, eds. OPEC Behavior and World Oil Prices, London, George Allen and Unwin, pp. 64 - 93.

US Department of Energy (DOE). (1994). Annual Energy Outlook, DOE/EIA-0383(84).

Chapter 5: World Hubbert Curve Oil Supply Forecast And The Reserve-Production Ratio Assumption*

ABSTRACT

The US Energy Information Administration (EIA 2000) has forecasted that world oil production will not peak until at least 2030 and maybe into the 22nd century. The EIA uses a Hubbert curve analysis to analyze oil markets by estimating an Ultimately Recoverable Reserve (URR) for the world of about 3 trillion barrels of oil and forecasting supply. The Hubbert analysis that the EIA uses is the same general analysis that M. King Hubbert (1962) used to forecast the US oil supply. However, careful analysis of world oil markets suggests that the world will not follow the same kind of Hubbert curve that the US followed. The US had a competitive oil market during all of the time that oil was produced in the US, but the world oil market is different. The world oil market does not have the same competitive characteristics as the US had, which may cause a higher oil reserve to oil production ratio for the world oil market than what occurred in the US.

One major assumption of the EIA's world oil supply is that the world will be able to attain a ten to one proven oil reserve to oil production ratio similar to

* "World Oil Supply Forecast and the Reserve/Production Ratio" by Douglas B. Reynolds, *2001: An Energy Odyssey*, Conference Proceedings of the 24th Annual International Conference of the International Association of Energy Economics (IAEE), April 25 –27 Houston, Texas, Copyright © IAEE 2001.

what the US oil supply market has been able to attain for many years now. This ten to one ratio is a crucial assumption to the EIA's forecast. However, an economic analysis of the exploration and extraction process of oil by single country oil companies within many OPEC and non-OPEC countries would suggest that the ten to one ratio is an incorrect assumption. Geologically and physically it is possible to attain a ten to one ratio of production to proven reserves, but that does not mean that economically that will happen. Reynolds (2000) shows that when a single country oil company, such as ARAMCO in Saudi Arabia or INOC in Iran or even PEMEX in Mexico, explores for and extracts oil, they tend to be more risk averse and consequently do not aggressively look for and develop new oil reservoirs as would happen within a competitive market such as the US has. This will cause world oil production to stay low in comparison to the US and this will change the Hubbert curve analysis.

If we assume that the world oil reserve-production ratio will be higher than the ten to one ratio that the US experienced, then oil production may never attain as high a production as the EIA estimates. One way to forecast what the maximum world oil production rate will be in the future is to use a modified Hubbert curve that is fitted to previous world oil consumption trend changes and the current estimated URR. World consumption peaked in 1973 and again in 1980. Those peaks may be used to mark where oil supply is at its maximum. Using those peaks and the EIA's estimated URR, I forecast a new Hubbert curve projection that incorporates the economics of risk averse oil producers. Using this analysis a worldwide price shock looks to be eminent in the next five to ten years or even sooner. Such a price shock occurs even if URR is 3 trillion barrels or even twice that.

INTRODUCTION

Ever since Campbell and Leherrere (1998) came out with their prediction of an oil shortage in the near future, other experts in the energy field have emphatically argued against such a conclusion. The US Energy Information

Administration (EIA) for example forecasts that oil production will not peak until at least 2030 and maybe into the 22nd century. Interestingly, both Campbell and Leherrere, and the EIA (2000) use a Hubbert curve analysis to analyze oil markets. M. King Hubbert invented the Hubbert logistics curve to accurately forecast the peak in US oil production, eight years before it occurred. Hubbert forecasted that US oil supplies would reach a peak in 1969 and decline thereafter. Actual US production peaked only one year later in 1970. Now however, there is renewed interest in using Hubbert's analysis to determine how large world oil supplies are and how long they will last. A consensus of expert opinion would suggest that world oil supplies will be plentiful for the next 25 to 100 years. However, there are a growing number of critics that suggest that oil supplies, using the Hubbert curve analysis, will soon peak and decline. In this paper, I will also use a Hubbert curve analysis to forecast oil supplies but I will add an extra term to account for the oil reserve to oil production ratio. It is clear that many oil-producing countries have a long history of high reserve-production ratios and yet no one is taking this into account in their analysis.

THE HUBBERT CURVE

In 1962 M. King Hubbert published his famous oil supply forecast for the US lower 48 states. Hubbert created a mathematical logistics curve, often called the Hubbert curve, that could be used to project future trends in oil discovery and production. Once the past trends for a given region is statistically fitted to a Hubbert logistics curve then the future discovery and production for that region can be forecasted as well as ultimately recoverable reserves of oil (URR). Hubbert used this curve to predict very accurately when US oil production would reach a peak and decline. Wiorkowski (1979), Ryan (1965), and Lynch (1994) have criticized Hubbert for not accounting for economic, technological and political changes in the oil market, all though Hubbert suggested that technological change is incorporated into his curve. Nevertheless, even with as much criticism as Hubbert received, his 1962 forecast for the peak in oil production for the US lower 48 was only off by one year. The forecast is extremely accurate considering, US lower 48 and indeed total US production has yet to go back up to that 1970 peak in production even with 30 years of new and improved technology, generally higher oil prices and de-regulation. New estimates for URR are also close to Hubbert's original estimate. Therefore, Hubbert's method can be very useful for forecasting when the peak in oil production for a given region will occur.

Cleveland (1991), Reynolds (1999), Slade (1982), and Uhler (1976) give theoretical reasons for why the Hubbert curve works. Cleveland and Kaufmann (1991), Moroney (1999), and Kaufmann (1991), incorporate economic principles into Hubbert's equations. Pesaran and Samiei (1995), Campbell and Leherrere (1998), Edwards (1997) and Campbell (1997) use Hubbert's equations to forecast oil supplies for the US and the world.

Lynch among others has criticized this type of forecasting because the estimated URR for the world and even the US has always changed over time, and has always gotten bigger. Therefore no one really knows what URR is. Still the US situation shows that it is possible with enough information to forecast URR fairly accurately and especially to forecast when production peaks for a given region. The most important thing to forecast is when production peaks. No one is predicting that US oil production will ever go higher than it went in 1970 even with better technology, higher prices and more de-regulation and even though the estimated URR for the US has increased. Therefore, it is not so much the actual URR that is important, but the estimate of when production will peak that is the most important.

One reason that Hubbert's curve is such an important tool for determining oil production potential for a given region is that it has been shown to determine an oil supply limit. In other words, the ability to supply oil can certainly be less than the Hubbert curve, but it cannot be greater. This conclusion stems from US evidence. In 1973, the price of oil went up close to four times higher, a 400% increase in price. Yet one year later, and even two, three or five years later, US oil extraction registered not even a 10% rise in production above the projected Hubbert curve. 400% to 10%, that is an elasticity of supply of about 0.025. It is about as close to perfect inelasticity as you can get.

Many oil experts claim that because of heavy regulations, it was not possible to explore for oil as adequately as was possible in the early to mid 1970's. Therefore, it is claimed that the US supply elasticity was inelastic merely because of regulation. However, in the early 1980's regulations were relaxed. Starting from 1979, the oil industry was deregulated and again there was a two fold increase in oil prices yet again there was barely a 10% increase in quantities above the Hubbert curve in the few years after the price shock. That is still an elasticity of supply at about 0.05, which is still extremely inelastic. Therefore, the Hubbert curve is an extremely inelastic supply limit. Once the world reaches that Hubbert curve limit, no matter how high the price goes up, supplies will not increase by much.

Campbell and Laherrere use a Hubbert curve analysis to estimate total world oil supplies. They predict a URR of 1.8 trillion barrels and a peak in oil production before 2005. If that is the case, then in less than five years, oil prices will increase to spectacular heights. An oil crisis of immense magnitude will ensue. However, even if the URR is much larger, we may still reach a Hubbert curve limit sooner than expected. Another determining factor for when the peak in production will occur besides the conventional Hubbert curve analysis itself is the reserve-production ratio.

The EIA has also made a world oil supply forecast using a Hubbert curve analysis. The EIA estimates a URR using geological data and scientific methods. They estimate a large medium and small URR. The EIA estimates the low value of URR at 2.25 trillion barrels of oil with a 5% chance that it will be less than that. It estimates the high value of URR at 3.9 trillion barrels of oil with a 5% chance that reserves will be greater than that. Its medium estimate is 3 trillion barrels. By tracking world oil consumption and using an estimate of 3 trillion barrels for URR, the EIA is able to forecast when the peak in world oil production will occur.

The EIA estimated future world oil demand using a low, high and medium rate of increase. These estimates are based on past economic performance of the world's economy and on projected world economic growth and the requirements for oil that such growth requires as well as on technological changes. The high estimate, for increasing world demand for oil, is a 3% increase in consumption per year. The medium estimate is 2% per year. The low estimate for oil demand is 0% per year, i.e. there will be no increase in oil consumption due to better technology and the greater use of oil substitutes. While a 0% projected growth rate in oil use does not seem tenable, given the large rebound in economic growth since the economic collapse of East Asia in 1998, at least it gives a lower bound. Finally, the EIA put together its projected Hubbert curve supply for world oil with its projected demand for oil. EIA uses its medium estimates for URR and its medium estimate for increases in oil demand to come up with its forecast for when oil production will reach a peak. Using these estimates, the EIA forecasts world oil supply to peak in 2037, as their best guess. They also make an alternative estimate that has oil supplies peak in 2030.

THE QUADRATIC HUBBERT CURVE

One of the problems with Hubbert's oil discovery and production logistics curve has been that it is time dependent. Because of this, if the rate of oil production goes down or even slows its increase, then it makes the path of production change substantially from Hubbert's time dependent logistics curve. Once oil production goes below the Hubbert's logistics curve it becomes difficult to track where the production limit is. An alternative Hubbert curve uses a simpler quadratic equation. This equation is derived by using the Hubbert time dependent oil production logistics curve and the time dependent cumulative oil production logistics curve and subsuming the time variable. The quadratic Hubbert curve is no longer time dependent but cumulative production dependent. The equation for the curve is:

$$QP = a \cdot CQP - (a/URR) \cdot CQP^2$$

where QP = Quantity of Oil Produced during each year; CQP = Cumulative Quantity of Oil Produced up to each year; URR = Ultimately Recoverable Reserves; a = a size parameter, which determines the height and width of the Hubbert curve

The new quadratic Hubbert curve has characteristics that make it easier to use. For example if actual oil production is below the quadratic Hubbert curve, it is easier to see where consumption is relative to the limits of supply and how far demand can increase before it reaches the Hubbert limit. Therefore, this new Hubbert curve is the supply limit. Putting both supply and consumption (demand) on the same graph will allow us to see how far away the Hubbert curve supply limit is from demand.

To see better how this works, consider Figure 5.1. In that figure is shown the EIA's projected Hubbert curve using their best URR estimate of 3 trillion barrels of recoverable oil in the world. The demand is also shown as a function of cumulative production. In this way we can easily see where the Hubbert curve limit is in relation to demand. The dates for all positions are given relative to consumption (Demand). For example in 1973, world cumulative oil production was at 328 billion barrels. The world oil consumption that year was about 21 billion barrels. The EIA's Hubbert curve puts supply potential for that year at 37 billion barrels or as little as 22.5 billion barrels per year for its alternative Hubbert supply curve. Therefore actual oil production for that year was below the large and the smaller Hubbert curve. The two EIA Hubbert supply curves depend on the variable "a" in the Hubbert equations.

As a further example consider 1983. In that year, cumulative world production was 549 billion barrels, actual consumption was about 22.2 billion barrels for that year, while the EIA's Hubbert supply potential was 52.2 billion barrels. For EIA's alternative Hubbert curve there was a supply potential of 34.5

billion barrels. Therefore, in 1983 there was a potential for the world to supply up to 52.2 billion barrels for that year. However, because demand was low, at 22.2 billion barrels per year, the world did not use its full potential. Since actual oil production was much lower than the Hubbert supply potential, then there were very competitive oil markets and this caused oil prices to fall. However, back in 1973, the difference between the Hubbert curve and actual supply was very tight and this may have caused large increases in oil prices.

Figure 5.1 EIA Forecasted Supply and Demand for World Crude Oil As a Function of Cumulative Production

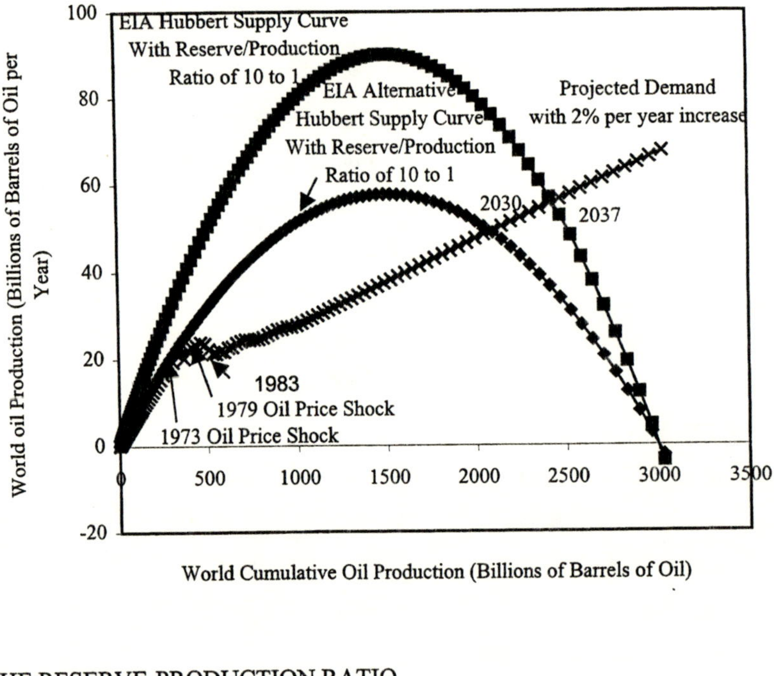

THE RESERVE-PRODUCTION RATIO

The term Ultimately Recoverable Reserves (URR) is for all past, present and future oil that will eventually be extracted. Since it is hard to know what future oil discoveries there will be, URR can only be estimated with a wide range of estimates. Another term in oil production however, is called proven reserves.

Proven reserves are the quantities of oil in the ground that are known beyond a shadow of a doubt to be producible. Proven reserves does not relate to the past or the distant future, it is only for reserves of oil that are known to exist and that will soon be extracted using existing technology. However, since every year some of those proven reserves are extracted and new oil discoveries are found, proven reserves always change from year to year. In the US it has almost always been the case that proven reserves amount to ten years worth of production at the current yearly rate of production. Thus there has always been ten years worth of proven reserves for one year's production, a reserve-production ratio of 10 to 1.

In the EIA report on world oil supplies, the authors correctly explain that US oil production always followed a Hubbert curve and that the reserve-production ratio for US oil production always stayed around 10 to 1 while it followed that Hubbert curve. In other words, there has always been ten years of proven oil reserves at the current years production rate, or one years oil production times ten usually equaled the proven oil reserves. The EIA further makes the assumption that if the US is able to sustain a 10 to 1 reserve-production ratio then the rest of the world can easily attain this ratio too.

A 10 to 1 reserve-production ratio is the industry rough standard for the limit on oil production. If an oil well or an oil field has say 10 million barrels of proven oil reserves then in one year at most only 1 million barrels can be extracted. That is the 10 to 1 ratio. If the oil producer tries to pump more than 1 million barrels of oil out in a year, say 2 million barrels, then it will have a 5 to 1 ratio. However, anytime the ratio goes lower than 10 to 1, that means a higher rate of extraction, and a higher rate of extraction can damage the oil field and actually reduce reserves in the long-run. Maybe this year the oil producer can extract 2 million barrels out of the oil field, but by reducing the reserve-production ratio, next year's production may go down to .5 million barrels and even lower after that. You lose oil with low ratios. Quicker production of oil causes oil to get trapped underground. Therefore, most oil fields never go below the 10 to 1 ratio.

On the other hand, many countries and oil fields do have reserve-production ratios greater than 10 to 1. Currently most of the major oil production nations have a reserve-production ratio much greater than 10 to 1. Mexico, and Nigeria have reserve-production ratios of about 30 to1. Saudi Arabia, Iraq, Iran, UAE, Kuwait have reserve-production ratios close to or in excess of 100 to 1. Salemah (1999), questions some of these OPEC oil producers claims for such high proven oil reserve estimates. However, even if we do not believe the official proven reserve estimates from each country, there is other evidence that oil production from a number of major oil producing countries is low, i.e. that those reserve-production ratios are high. Adeleman (1986) suggests that a comparison of the US to Saudi Arabia shows that there is much less oil well drilling, less exploration and development and less investment into the Saudi oil industry in general than the US even though oil reserves and potential oil production are greater in Saudi Arabia than in the US. In other words, there is more potential to find and extract oil in Saudi Arabia, then in the US and yet there is more investment in the US. That could be a reason for a high reserve-production ratio.

Reynolds (2000) suggests that the reason oil exploration and development investments are lower for some oil producer countries than in the US is due to risk aversion. These countries have a single country monopoly oil producing company where the country owns and controls all its own oil production. This causes the country to be much more risk averse to oil investment and consequently have lower oil production. Investments tend to be less aggressive and the pace of oil exploration and development is much slower than under a competitive environment. This however should not be interpreted as a bad thing. It is to the world's advantage that oil be conserved for the future. Oil is the most valuable energy commodity on Earth and always will be. Therefore, any market environment that conserves oil should be applauded.

Many OPEC countries and some non-OPEC countries have had reserve-production ratios of 50 to 1 or even as high as 100 to 1. Not only are these reserve-production ratios being currently maintained but they have been maintained for decades, even during the 1970's when oil prices were very high for over 12 years. This means that these countries rates of oil production will be much lower than a 10 to 1 ratio would suggest meaning that oil production will be much lower than what a conventional Hubbert curve analysis would suggest is possible.

ANALYSIS AND FORECAST

In Figure 5.1 is the EIA's latest world oil supply forecast using a Hubbert curve analysis and assuming that the world can increase oil production up to the point where there is a 10 to 1 reserve-production ratio. Since many OPEC and non-OPEC oil producing countries have never attained a 10 to 1 ratio and are likely never going to attain such a ratio, then it is wrong to make an assumption that the world will produce oil up to a 10 to 1 ratio. If the ratio is higher say 50 to 1 or even 100 to 1, then you can no longer use the usual Hubbert curve analysis.

For example, let us go back to the year 1973. In that year, world cumulative oil production was 328 billion barrels and oil production for that year was 21.2 billion barrels as stated above. The EIA's URR estimate means that there was a Hubbert curve supply potential of 37 billion barrels in 1973. So the question is, if the EIA's supply potential estimate is so large, then why was there an oil price shock in that year? Most oil analysts blame the 1973 oil price shocks on political causes rather than economic causes. However, in 1967 during the Israeli Six Day War, the Arabs tried an oil embargo that did not work, so why did such an embargo suddenly work so well in the 1973 Arab Israeli war? The reason is there was not a lot of extra oil production potential. In other words, there should have been plenty of excess oil supplies in 1973 according to the EIA's Hubbert curve. Therefore, there should not have been an oil price shock in that year. The fact that there was such a shock suggests that the world had a supply limit less than the EIA's Hubbert curve and that the world cannot easily attain a 10 to 1 reserve-production ratio.

Consider anther year, 1979. Again there was an oil shock. In that year, world cumulative oil production was at 460 billion barrels, and oil consumption for that year was 23.7 billion barrels. According to the EIA's URR estimate, there should have been a Hubbert curve oil supply potential for that year of a whopping 46.8 billion barrels or two times as much as was actually produced. Even their more conservative URR estimate shows a Hubbert curve potential supply of 30 billion barrels or 30% more capacity for 1979. Yet, there was an oil shock. The main reason given for this oil shock was the Iranian revolution, which cut off Iranian oil supplies. However, the cut in Iranian oil only lasted a few months and even with a cut in Saudi Arabian oil production at that time there should have been ample capacity to produce oil. The real reason for the price shock has to be because the world cannot attain the high Hubbert curve that the EIA forecasts, which may be because the world cannot attain the 10 to 1 reserve-production ratio the way the US could. Once oil demand hits against the lower supply potential, supply is very inelastic. That is why price went way up, but production did not increase.

OPEC oil producers tend to be much less aggressive in oil and gas exploration and development. They are conservative due to their natural political risk averse characteristics. It is this risk aversion that keeps oil production low and reserve-production ratios high. Therefore, we must assume that in fact a reserve-production ratio of 10 to 1 is impossible to attain worldwide and that it is not possible to produce oil at the EIA forecasted Hubbert curve.

An alternative analysis is to look at the world's potential Hubbert curve and then look at actual oil production, and then try to reconcile the differences. Consider Figure 5.1 again. Figure 5.1 shows the world's potential EIA Hubbert oil supply curve and actual consumption of oil in a quadratic Hubbert format, i.e. as a function of cumulative production. Notice that actual oil production is much less than the EIA Hubbert supply curve potential right now which should mean that OPEC would have a hard time keeping oil prices up. If we forecast oil demand at a 2% per year increase as the EIA does then demand crosses the

Hubbert curve supply at a cumulative quantity of 2.47 trillion barrels. That would occur in 2037, whereupon the Hubbert curve limit pushes production down at about 5% per year. That would suggest an oil price shock of tremendous magnitude in 2037.

The problem with using the EIA URR estimate and a US type Hubbert curve is that in fact right now OPEC is not having that hard of time keeping oil prices high, and the declines in consumption demanded in 1973 and 1979 seem to be awfully sudden. The sudden changes in demand in 1973 and 1979 occurred because oil prices suddenly shocked upward. It is however, surprising to find oil prices shocking up ward so suddenly when oil consumption was well below the Hubbert curve limit. Indeed, the very fact that oil prices suddenly, and not very temporarily, shocked upward suggests that the Hubbert curve at a 10 to 1 reserve-production ratio is not in fact the limit of oil production but that the Hubbert curve limit is much lower. Remember most oil producing countries in the world produce oil at a 50 to 1 or even a 100 to 1 reserve-production ratio, which is a rate of oil production 80% less than a 10 to 1 ratio. This means that a standard Hubbert curve should not be used to forecast world oil supply potential.

In Figure 5.2 is an alternative Hubbert curve called Scenario B. The Scenario B curve is created by finding a formula that fits the 1973 high point of oil production and the 1979 high point of oil production and the currently estimated URR. The equation used for this curve is

$$QP = [a \cdot CQP - (a/URR) \cdot CQP]^2 \cdot 0.78[(CQP/URR) + 1]^{-ex}$$

where ex = 2.5.

Other exponents for ex less than or greater than 2.5 do not fit the 1973 and 1979 high points as well. Scenario B assumes that the maximum world oil production is lower than what a 10 to 1 ratio would give. It was the Scenario B Hubbert curve that caused the 1973 and 1979 oil price shocks, although the second price shock was slightly lower than Scenario B due to Iran's slight reduction in production and Saudi Arabia's reductions thereafter. Scenario B suggests that demand will reach and exceed supply creating an oil price shock in the next five years, even with URR at three trillion barrels. If URR is even higher at say six trillion barrels it only delays the price shock by another five years. Therefore, we should not put our faith in expecting higher URR estimates to save us.

Figure 5.2 Forecasted OPEC Supply and World Oil Demand As a Function of Cumulative Production

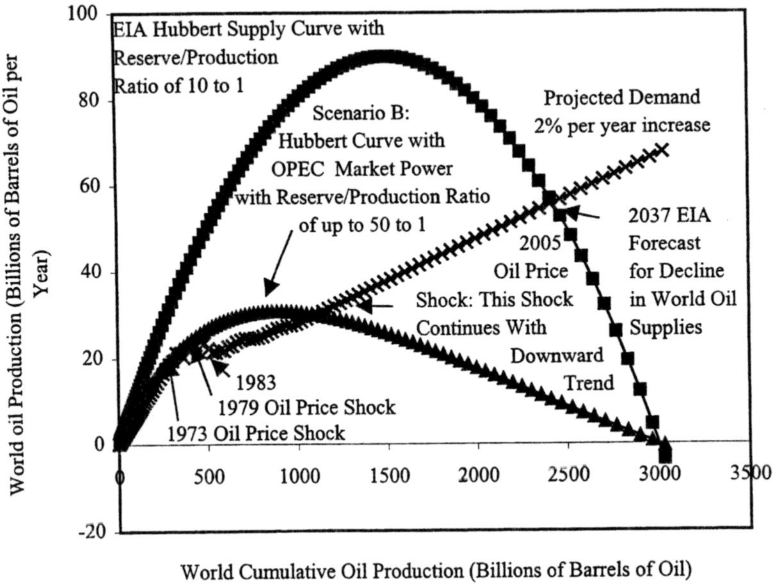

CONCLUSION

It is useful to use a Hubbert curve analysis to forecast oil supplies, but we should not assume that world oil production will be able to attain a 10 to 1 reserve-production ratio the way the US did. Oil producing countries with only one oil company controlling their production tend to be much more risk averse and do not aggressively explore for or develop oil. Their risk aversion and low production is a natural politically risk averse behavior. If a Hubbert curve with a high reserve-production ratio is used to forecast an oil supply limit using past production peaks, then it is possible that we will have an oil price shock in less than 5 years. The US EIA is doing the US and the world a disservice by forecasting that there will not be any peak in oil production until decades into the future. They have not adequately characterized the real world economics in their models. By assuming a uniquely American oil exploration and development experience, which is far removed from the supply economics of the rest of the world, they are not preparing the world or the US for potential oil shortages in the very near future.

Furthermore, because the US experience gives conclusive evidence that a 10 to 1 reserve-production ratio is possible, the US Department of Energy will continue to be tempted to cajole OPEC countries, especially Saudi Arabia, to get their production higher. Such cajoling and pressure by the US on OPEC and Saudi Arabia is not likely to ever produce greater oil supplies. OPEC countries are naturally politically risk averse when it comes to business investment. They are naturally politically risk averse because they have to be receptive to what their constituents want which is not to make risky investments and not to make mistakes, especially in regards to new oil exploration and development. One can no easier change OPEC countries risk aversion then change the Hubbert curve itself.

Finally, the US and the West need to applaud OPEC efforts to conserve oil. Oil is the most valuable energy commodity on earth. We have had over 100 years of technological development since the first large scale oil use started and in those 100 years of technological development, we have yet to find an adequate substitute for oil. This is not likely to change in the next five to ten years. Keeping oil prices low now, may help the economy now, but it will not help the economy prepare for the future.

REFERENCES

Adelman, M.A., (1986). "Scarcity and World Oil Prices," The Review of Economics and Statistics, Volume 68, Number 3, August pp. 387 – 397.

Campbell, C. J., (1997). The Coming Oil Crisis, Multi-Science Publishing Company & Petroconsultants S.A,.

___________, and Jean H. Laherrere, (1998). "The End of Cheap Oil," March Scientific American, pp. 78 - 83.

___________, (1998). response to letters, The National Interest, Summer, pp. 116, 117.

Cleveland, Cutler J. (1991). "Physical and Economic Aspects of Resource Quality, The Cost of Oil Supply in the lower 48 United States, 1936 - 1988" Resources and Energy, Volume 13, pp. 163 - 188.

______________ and Robert K. Kaufmann (1991). " Forecasting Ultimate Oil Resources and its Rate of Production: Incorporating Economic Forces in the Models of M. King Hubbert" The Energy Journal Volume 12, Number 2, pp. 17 - 46.

Edwards, John D., "Crude Oil and Alternate Energy Production Forecasts for the Twenty-First Century: The End of the Hydrocarbon Era," AAPG Bulletin, August 1997.

Hubbert, M.K. (1962). Energy Resources, A Report to the Committee on Natural Resources: National Academy of Sciences, National Research Council, Publication 1000-D, Washington, D.C., pp. 54, 61, 67.

Kaufmann, Robert K. (1991). "Oil Production in the lower 48 States: Reconciling Curve Fitting and Econometric Models" Resources and Energy, Volume 13, pp. 111 - 127.

Lynch, Michael C., (1994) "Bias and Theoretical Error in Long-Term Oil Market Forecasting," in Advances in the Economics of Energy and Natural Resources, John R. Moroney, ed., JAI Press.

Moroney, John, and Doug Berg. (1999). "An Integrated Model of Oil Production," The Energy Journal, Volume 20, Number 1, pp. 105 – 124.

Pesaran, M.H. and H. Samiei (1995). "Forecasting Ultimate Resource Recovery," International Journal of Forecasting, Volume 11, Number 4, pp. 543 - 555.

Reynolds, Douglas B. (2000) "Modeling OPEC Behavior: Theories of Risk Aversion for Oil Producer Decisions" *Energy Policy*, Volume 27, pp. 901-912.

_________. (1999) "The Mineral Economy: How Prices and Costs Can Falsely Signal Decreasing Scarcity" *Ecological Economics*, Volume 31, Number 1, pp. 155-166.

Ryan, J.M. (1965). "National Academy of Sciences Report on Energy Resources: Discussion of Limitations of Logistic Projections," Bulletin of the American Association of Petroleum Geologists, Volume 49, Number 10 pp. 1713 - 1720.

Salameh, Mamdouh G. (1999). "Technology, Oil Reserve Depletion and the Myth of the Reserve-to-Production Ratio," OPEC Review, Volume 23, Issue 2, June, pp. 113 – 125.

Slade, M. (1982). "Trends in Natural Resource Commodities Prices: An analysis of the Time Domain" Journal of Environmental Economics and Management, Volume 9, pp. 122.

Uhler, R.S. (1976). "Costs and Supply in Petroleum Exploration: The Case of Alberta," Canadian Journal of Economics, Volume 9, Number 1, pp. 72 - 90.

United States Energy Information Agency, (2000) "Long Term World Oil Supply," EIA web site, EIA, 7/28/2000, http://www.eia.doe.gov/pub/oil_gas/petroleum/presentations/2000/long_term_supply/index.htm.

Wiorkowski, J.J. (1981). "Estimating volumes of remaining fossil fuel resources: a critical review," Journal of the American Statistical Association, Volume 76, Number 375, September, pp. 534 -547.

Chapter 6: Energy Grades and Historic Economic Growth*

INTRODUCTION

In 1709, Abraham Darby invented the coking process, which led to the use of coal in 18th century England. From an economic stand point, one could say that this event more than any other ushered in the industrial revolution with its dependence on coal and steal produced from coal. However from an engineering perspective, there is another cause of the industrial revolution that is subtler. This cause is the physical make up of the energy resources available to England.

According to Simon Kuznets, an economic epoch, which is a period of time defined by rapid population increase for a given region, "is determined and shaped by the application and ramification of an 'epochal innovation'." i.e. new significant technologies. [1]

Rondo Cameron further states,

> A possible explanation for the correlation of population growth/stagnation/decline with income movements can be fashioned by analyzing the interaction of the fundamental determinants of economic development (land, labor, capital and entrepreneurial capacity). With a

* "Energy Grades and Economic Growth," by Douglas B. Reynolds, *Journal of Energy and Development*, Spring 1994, Volume 19, Number 2, pp. 245-264,

> given technology, the resources available to a society set the upper limits to its economic achievements ... technological change by increasing productivity and opening up new resources has the effect of raising the ceiling. [2]

This emphasizes technology as the major ingredient for periods of high economic and population growth. However, I believe another ingredient, equally as important as technology, is the grade or inherent value of energy resource inputs available to an economy. This has to do with productivity. Each type of energy resource has an inherent physical potential for being more or less productive and that potential is the energy grade. Higher-grade energy resources have more potential for being productive than lower-grade energy resources.

Energy is the driving force behind industrial production and is indeed the driving force behind any economic activity. However, if an economy's available energy resources have low-grades, i.e. low potential productivity, then new technology will not be able to stimulate economic growth as much. On the other hand, high-grade energy resources could magnify the effect of technology and create tremendous economic growth. High-grade resources can act as magnifiers of technology, but low-grade resources can dampen the forcefulness of new technology. This leads to the conclusion that it is important to emphasize the role of the inherent nature of resources in economic growth more fully.

To see better how this very subtle idea is a not so subtle cause of the industrial revolution, and possibly other economic epochs, we must look at some simple physics of energy resource characteristics. I believe that the most important resources for economic achievements are energy resources, therefore, I look at ways to compare energy resources.

THE ENERGY RESOURCE CHARACTERISTIC GRADE

In order to understand why some energy resources are better than others, we need a way to compare them. One way to compare energy resources is the energy grade concept defined here. This concept identifies the physical characteristics of competing energy resources that allow the economy to more cheaply extract services from each BTU [3] of energy. There are four grades.

1. Weight Grade (BTU / lb.)
2. Volume Grade (BTU / cubic foot)
3. Area Grade (BTU / acre)
4. State Grade (Liquid, Gas, Solid, Field)

Consider these grades in detail:

1. Weight Grade

The weight grade determines how much energy there is per pound of energy resource. For example, coal has about 12.7 thousand BTU/lb, natural gas about 10 thousand BTU/lb, oil about 19.3 thousand BTU/lb, and an electric battery typically has 100 BTU/lb. Electric batteries then are very heavy compared to their energy output which is why electric cars do not have very good driving ranges.

The weight grade determines energy performance. Usually, transportation devices must carry their fuel source along with them during use. The lighter is the weight of the fuel they use, the less energy they require to carry that fuel around which is why consumers and producers will be willing to pay a premium for higher weight grade energy resources.

2. Volume Grade

The volume grade determines how much energy there is per unit of volume of the energy resource. Natural gas is very bulky at about one thousand BTU/cubic foot at standard atmosphere and pressure, and 177 thousand BTU/cubic foot at 3000 psi. Oil, though, has about one million BTU/cubic foot. The volume grade is important again, because it determines performance for certain energy uses. For example, if we had to use natural gas in place of oil for cars, the volume of the fuel tank would have to be much bigger and thus much heavier, or if it was the same size, then refueling would need to be done more often.

A low volume grade energy resource is also difficult to transport. For example, a low volume grade resource like natural gas can be many times as expensive to obtain from an over seas source, such as the Middle East, than from the North American continent due to storage expense during transportation. So here again consumers and producers will be willing to pay a premium for higher volume grade energy resources.

3. Area Grade

The area grade determines how much energy there is per area of occurrence of the energy resource in its original state, i.e. how much energy per acre. For example, the area grade of wood is roughly 1 to 5 Billion BTU/acre because wood is spread out in forests over many acres. Its original energy state, then, is much more spread out. The area grade for oil is usually tens or hundreds of billions of BTU/acre, as it is found in thick under ground reservoirs in a high volume grade state.

The area grade determines how much service including cost savings the economy can extract from a given energy resource. If the energy content of the resource is spread out, then it costs more to obtain the energy, because a firm has to use highly mobile extraction capital, which must be smaller and so cannot enjoy increasing returns to scale. If the energy is concentrated, then it costs less to obtain because a firm can use larger scale immobile capital that can capture increasing returns to scale. Therefore, energy producers will be willing to pay an extra premium for higher area grade energy resources.

4. State Grade

The state grade defines what form or state the energy resource occurs in. The four major state grades are the following:

1.) Liquid
2.) Gas
3.) Solid
4.) Field

1.) Liquid

The liquid state grade is simply where the energy resource occurs in a liquid form at standard atmosphere and pressure, such as oil does. This state is the highest state grade, because energy resources that are liquids are easier to transport and use than any other energy state. For example, a machine can use less moving parts to inject, burn and remove a liquid in a burning chamber, such as a piston cylinder, than it can a solid. Less moving parts usually means less costs. Also one can more easily transport and store a liquid, than a solid or a gas, since a producer can carry a liquid in un-pressurized containers or pump it through pipes. This makes liquids cheaper to use than energy resources that occur in other states.

2.) Gas

Gas at standard atmosphere and pressure is the next highest state grade. A gas is more difficult to transport and use than a liquid, because it by nature must have a lower volume grade and must be kept under pressure. However it is still fairly easy to use. A machine can inject gas into a burning chamber just like a liquid.

3.) Solid

A solid energy resource is the third highest state grade. It is simply an energy resource in solid form at standard atmosphere and pressure such as coal or wood. Solid fuels are more difficult and thus more costly to use, because in order to burn them, complicated mechanisms must continually place them in a burning chamber and remove the ashes once they are burned. A machine cannot pump the fuel into place but must mechanically move it. This makes solid fuels more costly to use per BTU than liquid or gas fuels in many energy uses.

4.) Field

A field energy resource includes such phenomena as radiation fields, like solar and nuclear power, and pressure fields, like wind energy and hydropower and are the least useful state grade.

The main problem with fields is that they are difficult to store. For example, one of the biggest problems with solar energy is storing the daytime heat energy for nighttime use. The only field that does not have this problem is nuclear fields, but they are difficult to contain from creating environmental hazards.

There are four different physical manifestations of the field state grade:

(1) Pressure
(2) Electric
(3) Magnetic
(4) Radiative

(1) Pressure

The pressure field is where there is a difference in pressure. An example of such a field is like the wind pushing a windmill. The wind creates a pressure differential that in turn pushes the arms of the mill. Another such field occurs with hydropower. Firms often use pressure fields to make electricity.

(2) Electric and (3) Magnetic

Electric and magnetic fields are most often used in energy conversion processes only after another energy resource is used. Therefore, they are not sources of energy.

(4) Radiative

A radiative field is like solar energy. The sun's light radiates on the Earth. That radiative energy in turn creates heat or even electricity. All nuclear power too is a field type characteristic grade, since nuclear fuels, like Uranium, radiate alpha, beta or gamma particles.

Table 6.1. Energy Grades
(In British Thermal Units)

Energy Resource	Weight Grade (Btu/pound)	Volume Grade (Btu/Cubic Foot)	Area Grade (Btu/Acre)	State Grade
Oil	20,000	1 million	10 – 1000 billion	Liquid
Natural Gas	24,000	1000 at std [a]; 177000 at 3000 pounds per square inch	10 – 1000 billion	Gas
Coal	8,000 – 13,000	650,000 – 1 million	10 – 1000 billion	Solid
Oil Shale	Up to 2,000	200,000	10 – 1000 billion	Solid
Uranium	B	B	trillions	Field
Wood	8,000 dry	250,000	1 – 5 billion	Solid
Ethanol	11,500	560,000	Made from grain	Liquid
Methanol	8,600	425,000	Made from natural gas	Liquid
Hydrogen	63,500	325 at std [a]; 120,000 when super cooled	Made from electricity or coal	Gas; Liquid when super cooled
Solar Energy	b	b	1 – 10 million every hour	Field
Grain	7,000 dry	200,000	3 – 40 million every year	Solid
Wild Game	7,000	350,000	Up to 300,000 every year	Solid
Wild Berries/ Nuts	Up to 7,000	10,000 to 400,000	Up to 3 million every year	Solid

[a] std is standard atmospheric temperature and pressure; [b] Not applicable because it is a field grade

ENERGY RESOURCES AND THE HISTORY OF ECONOMIC GROWTH

In Table 6.1 and Figures 6.1, 6.2, 6.3 and 6.4, there is listed and shown many energy resources and their characteristic grade values. These show how the different energy resources compare with each other. Note that oil is one of the most valuable energy resources since it leads the other resources in most categories.

Figure 6.1. Energy Sources State Grades
(at standard atmosphere and pressure)

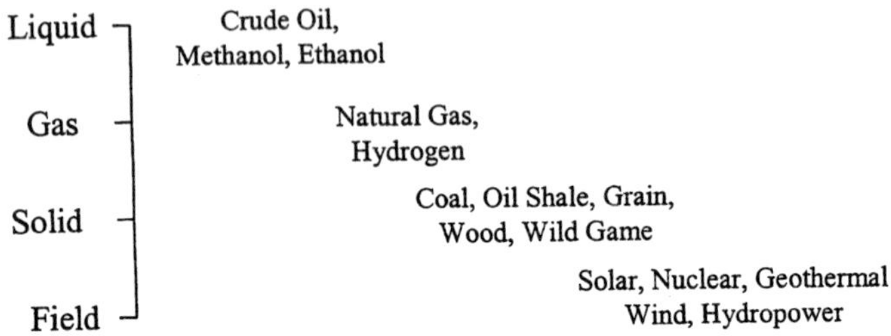

Figure 6.2. Energy Sources Ranges of Area Grades
(in billion British thermal units per acre as found in nature)

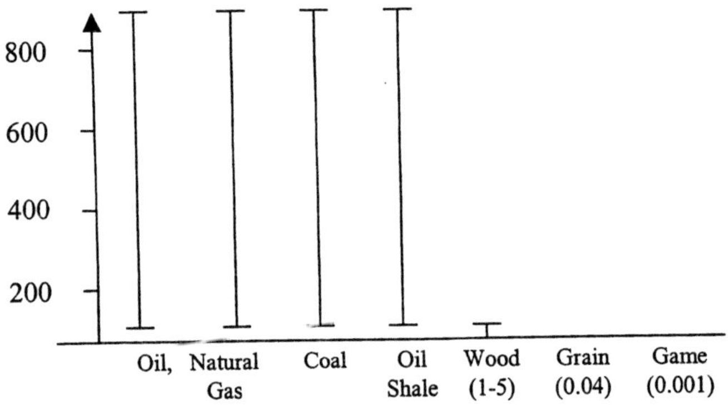

This comparison gives us a new way to analyze economic history. Many of the greatest economic epochs in history seem to occur at the point in time when the economy starts to use a high-grade energy resource. Examples are ancient man's switch from hunting to farming, which created the great ancient civilizations of Egypt and Mesopotamia; England's switch from wood to coal in the 18th century, which helped to create the industrial revolution; and the US's switch from coal to oil in the 20th century, which created the modern mobility revolution, also identified by some historians as the second industrial revolution [4]. All of these were changes to higher-grade resources.

Figure 6.3. Energy Sources Volume Grades
(in thousand British thermal units per cubic foot at room temperature)

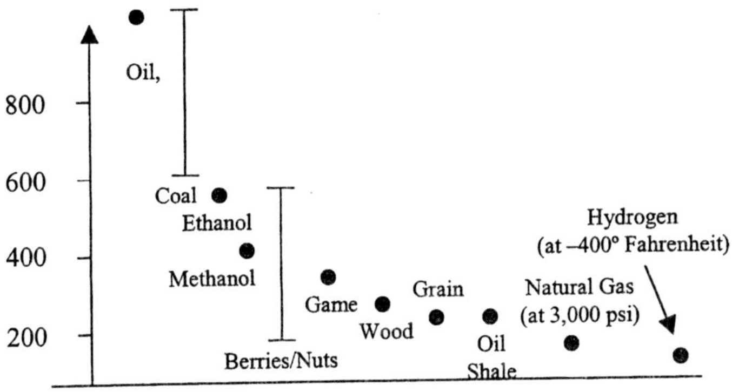

Figure 6.4. Energy Sources Weight Grades
(in thousand British thermal units per pound)

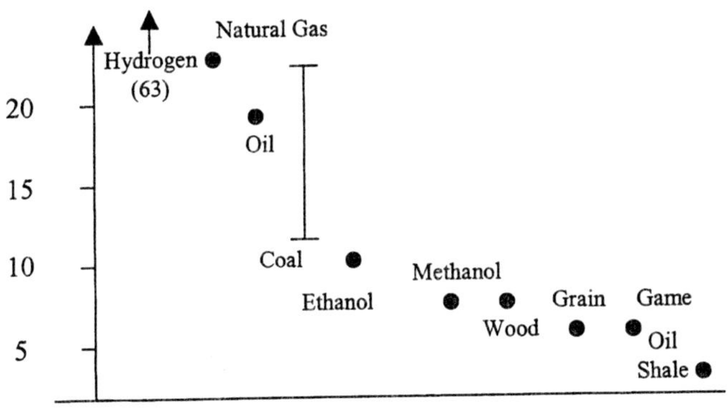

Consider Figures 6.1 – 6.4 in these terms. In Figures 6.5 – 6.8 are shown the same grade characteristics of Figure 6.1 – 6.4 only this time organized chronologically by when humans used the resource. From these grades, we can infer that as humans have advanced over time, they have used higher-grade energy resources. I believe that one of the causes of human economic development is the fact that humans used higher-grade energy resources, which created lower costs for production. Furthermore, I believe that much of economic growth is not due to better technology alone, but rather due to a combination of technology and higher-grade resources.

Figure 6.5. State Grade History
10 million B.C. to 2000 A.D.
(at standard atmosphere and pressure)

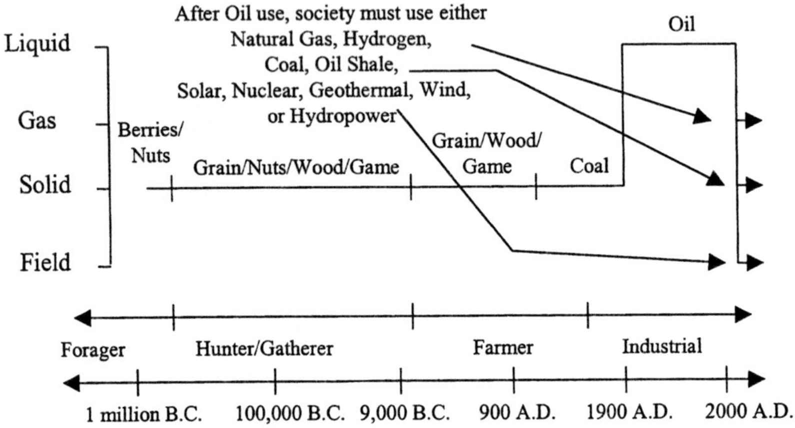

Figure 6.6. Area Grade History
10 million B.C. to 2000 A.D.
(in billion British thermal units per acre as found in natur)

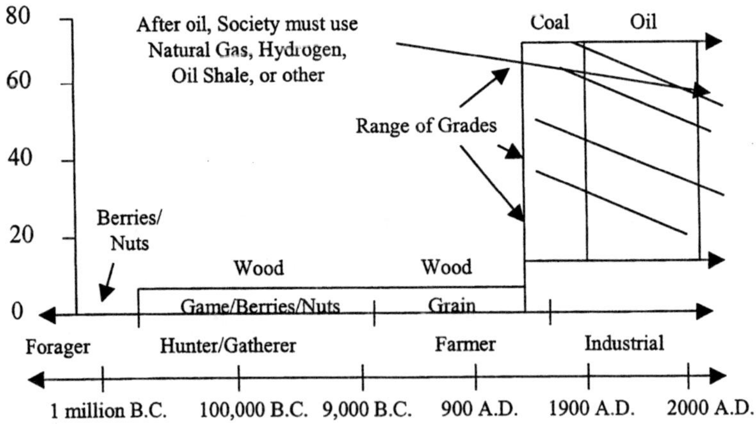

It can be said that technology was the reason for the use of higher-grade resources, and thus it is technology that is the only reason for human economic growth. This is true. However advances in technology without the availability of higher-grade resources would surely not have created as much economic growth success as was possible with the availability of these higher-grade resources. This gives some evidence that the degree of success in human economic growth was determined to a large degree by the grade of energy resources available. Consider the example of England's energy switch in the 18th century.

Figure 6.7. Volume Grade History
10 million B.C. to 2000 A.D.
(in thousand British thermal units per cubic foot at room temperature)

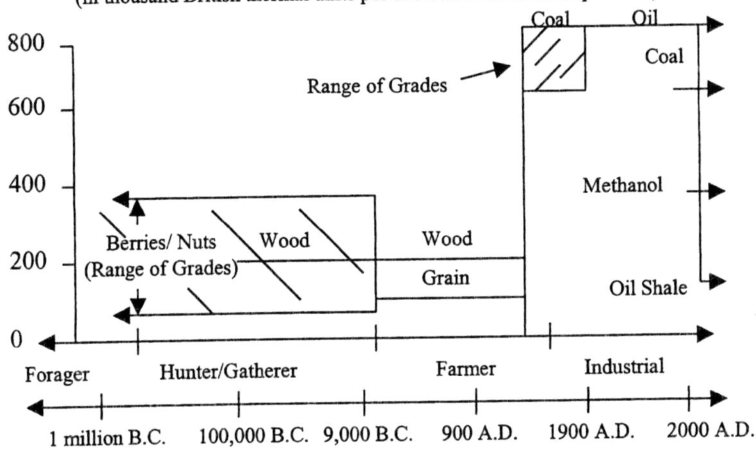

18TH CENTURY ENGLAND'S CHANGE FROM WOOD TO COAL

To show how the grade level of different energy resources affects the overall cost of using energy, consider England's energy resource switch in the 1700's from wood to coal. England needed wood throughout the Middle Ages for fuel and building. However, by the 18th century, England's forests were very depleted, and it threatened England's economy. [5] The price index of wood charcoal quadrupled in 100 years from 1560 to 1660 while the price index of everything else doubled in the same time span. This signaled problems with getting enough energy. [6] Eventually though, coal replaced wood for energy.

Consider this transition in detail. The area grade for wood is roughly 1 to 5 billion BTU's per acre but coal can have 10 billion to 1 trillion BTU's per acre. In fact, it often averages about 50 billion BTU's per acre or more. Therefore, coal's area grade is about 10 times greater than wood's. This allows miners to set up and use large scale production machinery, because they do not have to move the machinery from coal bed to coal bed since one coal seem has a lot of energy per acre. This is in contrast to wood charcoal where the gathering of wood required relatively more mobile capital. The result was that coal mining could have increasing returns to scale. Even such archaic mining machinery as they had in the 18th century benefited from that, which was why coal at that time was cheaper.

Figure 6.8. Weight Grade History
10 million B.C. to 2000 A.D.
(in thousand British thermal units per pound)

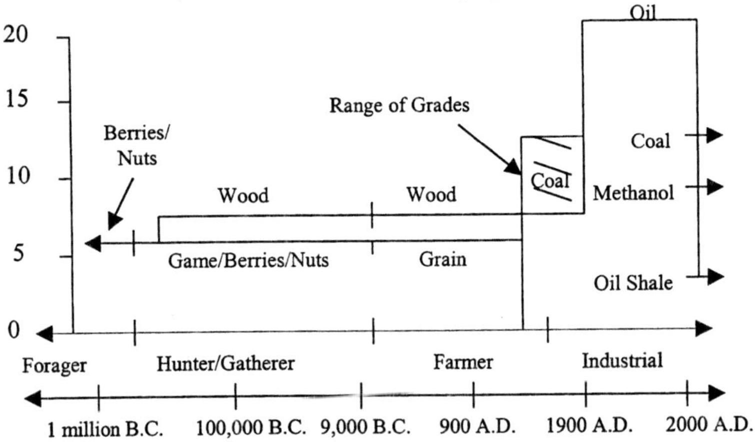

Since fuel is the largest input for making iron, iron furnaces were located closer to the fuel source than the iron ore. So it was the fuel source that determined the economies of scale of iron making. With coal as the new source of fuel for iron, producers could set up larger and more iron production furnaces close to a single coal seam. When iron production depended on wood, which was expensive to transport, wood being half the weight grade as coal, then that meant that the size of an iron furnace could only be as big as the supply of wood would allow. So charcoal furnaces had to be kept smaller and more spread out which caused lower returns to scale and so higher costs. Ashton gives some examples,

> A single furnace associated with a single forge was clearly the predominant order (before 1750). ... In 1549, there were 23 men working a furnace at Sheffield (Sussex) in addition to the two wainmen who attended the fourteen oxen and at a forge in the forest of Worth 33 men were engaged. ... at Duffield in 1691 it appears that 105 tons of metal were cast in 18 weeks, and that 75 tons of this pig iron were required to make the 50 tons of bar iron, which were produced at the forge in six months. [7]

So there were comparatively few laborers and low outputs.

However these furnaces often needed a lot of extra labor just to get the wood fuel. Again Ashton says,

> ... at Backbarrow (using the method of wood charcoal casting) ... there were in 1714 no fewer than 130 people supplying fuel to the works - sometimes in almost minute quantities. [8]

However a few decades after Abraham Darby invented coke, furnaces began to be located near coalmines. The furnaces were bigger, there were more of them located in one place, and outputs were higher. Ashton states, "...by 1803 Richard Crausby owned ... six furnaces and employed over 2000 men at Cyfarthfa." [9] Most furnaces in that day produced 40 tons of iron per week, which is about a seven-fold increase per furnace. [10] So with coal it was possible to have larger operations closer together creating increasing returns to scale. Plus the coalmine operations themselves could be bigger and enjoy increasing returns to scale.

However, in addition to having the larger returns to scale, the operations of mining and iron production were all in one place, which allowed for more specialization and also greater technical interaction, which created new technological advancements. As Raistrick says,

> ... it was now (from 1760 onward) an economic proposition to apply a large cylindered (watts) engine to mine pumping. This in turn made much deeper and more extensive mining possible and a cycle of development by interaction - foundry - engine parts - deeper and better pumping - easier and cheaper ore and fuel - larger furnaces and foundries - larger engines - was soon established. [11]

So we see that larger scale operations with more specialization and greater technological progress was a direct result of the high area grade of coal, because more operations were located closer together. If England ran out of wood and had to use a lower area grade fuel for iron production such as grain turned into alcohol or renewable forests, then surely England could never have had the economies of scale and so the technological leap that it did with high area grade coal. If England did continue to use wood however, it may have required the use of farmlands so that it could not support as many of its citizens as it had. This leads us to conclude that there would not have been as large a bang in the industrial revolution, nor may never have been an industrial revolution, without a high area grade resource like coal. Furthermore, we cannot see technology as being the only ingredient for an economy being able to overcome a resource shortage. Rather, it is technology applied to a new abundant higher-grade resource that more often than not saves an economy hit by a resource shortage.

Another way to view the importance of high-grade resources is to ask, what if England of the 1700's had today's technology but still only had wood, grains, wind and solar energy as its primary energy resources, without even the coal it used for heating, would the economic growth for England of the last 200 years have been possible?

The answer is that even today it is more expensive to use these alternative energy resources then current high-grade energy resources of coal, oil and natural gas. The fact that England's economy, and all industrial economies, choose not to use those alternative energy resources during the oil price shocks shows that it is more efficient to use the high-grade ones. This implies a loss of GDP if our economies were forced to use the lower-grade resources, which further implies England's growth of the last 200 years was greater with the high-grade resources than without them. The magnitude of the impact of having a high-grade resource available is not possible to find.

AN ENERGY THEORY OF VALUE

At this point one might ask if it is possible to compare competing energy resources on a price per BTU basis. Such a price per BTU criteria would allow us to have a better basis for comparing past energy transitions and future energy transitions. Unfortunately, such a criteria does not work as a comparison because competing energy resources have different characteristics that BTU content cannot capture. For example, consider the following problems.

An electric car using electricity at $14/MMBTU is less costly to run per mile even with maintenance costs than a regular gasoline car using gasoline at $8.00/MMBTU, [12] yet most people drive gasoline vehicles rather than electric vehicles. The reason for this is the overall service of gasoline vehicles. Gasoline vehicles have a range of 200 to 400 miles or more before needing to be refueled where as electrics can only go 30 to 60 miles. Plus it takes five minutes to refuel a gasoline car but an electric requires 30 minutes to 8 hours of recharging depending on the system before it is ready to go again. So electrics are inconvenient. Thus the price per BTU does not account for the difference in service provided and so cannot take into account consumer preferences nor total producer costs of using such alternative fuels which in turn will decide which energy resource is best.

Another problem with the cost per BTU concept is that when energy resources are converted from one form to another, there is typically a 10% to 90% reduction in energy even while there is a cost associated with such conversions. Thus natural gas at a well head can cost as little as $0.10/MMBTU, but when it is converted into methanol it will cost about \$8.00/MMBTU and there will be a loss of 40% of the original energy content causing greater scarcity of the natural gas. [13] However, this change in cost may be worthwhile, because the change in energy characteristics from natural gas to methanol may be worth the extra cost. Nevertheless, the cost or price per BTU concept does not capture that added value gained by turning natural gas into methanol nor does it explain the higher loss of the natural gas source.

Another problem is that location of energy can change the price per BTU. Natural gas for example that is produced in say Saudi Arabia would cost at least \$4.00/MMBTU delivered in New York city where as natural gas from Pennsylvania can cost as little as \$0.35/MMBTU in New York [14]. The reason for such a huge difference is that natural gas from Saudi Arabia must be shipped in cryogenic tankers that cool the gas to a super cooled liquid state in order to minimize the cost of transportation. The energy required to get the gas that cold and keep the gas super cooled during the duration of transit costs so much that it causes the price of the gas to be more costly for such long distances even though that is the cheapest method of transport. Thus the location of an energy resource affects the price per BTU. Then the question is which price does one use, the price of origin of the gas or the price of destination, and if we use the price of destination then do we use the price of gas coming from Pennsylvania or the price coming from Saudi Arabia.

Given these inadequacies, the price per BTU of energy does not adequately capture the real value of competing energy resources and so cannot determine which energy resources are most competitive. This means we cannot simply compare energy resources on a price per BTU basis nor even a simple BTU basis, but must compare them on a grade basis. Further, when energy statistics are presented in BTU terms and not grade terms, then those statistics tacitly assume one to one substitution per BTU between energy resources of different grades. This is clearly is not the case. I recommend that energy statistics for supply and demand of different types of energy not be lumped together using the BTU measure.

FUTURE ENERGY TRANSITIONS

If we compare our emerging energy transition of oil to oil alternatives with other energy transitions in history, it is important to distinguish the change in the grade level of the competing energy resources from changes in technology. Low-grade energy resources create higher cost production than high-grade resources, which in turn produces a drag on the economy. This leads to the conclusion that humans have made several energy transitions before in history and enjoyed growing economies during or after these transitions, but that most of the more successful energy transitions in history were transitions to higher-grade energy resources not to lower ones. Therefore, I am concerned with how successful the next energy transition will be. I propose three alternative scenarios for the future transition from oil to oil alternatives.

1) The economy goes to a higher-grade resource, creating a successful energy transition.
2) The economy goes to a lower-grade energy resource with better technology, creating a less successful but palatable energy transition.
3) The economy goes to lower-grade energy resources with virtually unchanged technology, creating an unsuccessful energy transition.

The first scenario suggests we go to a higher-grade energy resource. However, the question is, what other high-grade resources exist. Most alternative energy resources such as natural gas have mostly lower-grade characteristics. The only alternative energy resource that might be of higher-grade is solar energy or nuclear energy resources, all of which are fields. Solar energy is hard to store. This makes solar energy impractical to substitute for oil for transportation and other purposes. Nuclear energy has other problems.

Nuclear fusion energy, which is the energy of the sun, uses water and would have an extremely high area, weight and volume grade. However, fusion is only possible in large-scale facilities with multi million dollar lazars. It can therefore only be used for producing electricity on a large scale. Because of the highly technical and specialized capital, materials and labor it needs, it does not look to be any cheaper an energy resource for producing electricity than coal is currently. Nuclear fission energy, or conventional nuclear power, also has a high area, volume and weight grade for its energy source of Uranium. However, Uranium is extremely toxic and difficult to keep contained from the environment, and unless breeder reactors are used, a highly dangerous proposition, then uranium supplies will shortly run low. Furthermore, conventional nuclear power is too unsafe for using in numerous small-scale operations such as running trucks.

The second scenario suggests we go to lower-grade resources but with improved technology. This is like how Western Civilization has achieved greater productivity in farm production over time due to new technology even though soils are the same. If technology does advance fast enough, than maybe the negative effects on the economy of going to a lower-grade energy resource will be minimized.

The third scenario suggests the worst possible out come, that the world's economies will endure an unsuccessful energy transition and have very low growth rates or even economic decline. As the economies go from the high-grade resource of oil to lower-grade resources, there could be drag on economic productivity. Whether technology will or will not come through for us is very open to debate.

CONCLUDING REMARKS

I believe that many economic epochs in history should not be defined as being caused mostly by epochal innovations, but rather by innovations as well as the change in energy grade to higher-grade energy resources. Furthermore, both technology and high-grade energy inputs deserve equal status as causing economic epochs. The greatest economic epochs seem to occur when there are energy transitions from low-grade to high-grade energy resources. The economic intuition behind this is that higher-grade energy resources allow for increasing returns to scale since they are either less spread out, more concentrated when in bulk, or more flexible in use.

In our own day, we must eventually move to lower-grade energy resources as we slowly run out of oil. Therefore, we might expect the transition from oil to oil alternatives to be a decisively less successful energy transition than previous energy transitions in history, since all the previous transitions were from low-grade to high-grade energy resources, and the coming oil transition is from a high energy resource of oil to lower-grade energy resources. Greater technical progress should help our impending energy transition, but certainly we need to expect a lower growth rate during and probably after the next energy transition. Since industrialized country growth rates seem to be lower after 1973, the year oil production limits became apparent with high prices, then this could mark the beginning of lower growth rates due to the world's economies having to transition to lower-grade energy resources with their corresponding higher costs of production and lower productivity for the economy.

NOTES

[1] Cameron, Rondo. (1989) A Concise Economic History of the World. Cambridge University Press, p.187.

[2] Ibid., p. 9.

[3] The term British thermal unit is the energy required to heat 1 gallon of water by 1 degree Fahrenheit.

[4] Barnes, Harry Elmer. (1942) An Economic History of the Western World. New York: Harcourt, Brace & Co., p. 445.

[5] DeVries, John. (1976) The Economy of Europe in an Age of Crisis: 1600 - 1750. Cambridge University Press, pp. 166 -167.

[6] Cipolla, Carlo M.. (1976) Before The Industrial Revolution: European Society and Economy, 1000 - 1700. W.W. Norton and Company Inc., N.Y., pp. 266, 268.

[7] Ashton, Thomas Southeliffe. (1924) Iron and Steel in The industrial Revolution Manchester University Press, p. 96.

[8] Ibid. p. 187

[9] Ibid. p. 96

[10] Ibid. pp. 6-7.

[11] Raistrick, A.. (1933) Dynasty of Ironfounders London 1933 p.148, in Alan Birch The Economic History of The British Iron and Steel Industry. Franck Cass and Company Limited, London, p. 59.

[12] Starr, Gary. (1991) The Shocking Truth About Electric Cars. Earth Options, Santa Barbara, California, January.

[13] Othmer, Donald F.. (1982) "Methanol is the Best Way to Bring Alaska Gas to Market". Oil and Gas Journal. November 1, p. 84.

[14] International Energy Agency. (1982) Natural Gas: Prospects to 2000. Paris.

Chapter 7: Entropy Subsidies*

INTRODUCTION

During the oil price shocks of the 1970's, there were also price increases for other energy resources, most notably oil from oil shale. One reason for the increase in shale oil costs is a phenomenon where by a high quality or high-grade energy resource like oil creates lower costs for alternative fuels. When the high-grade energy is less available, it is less effective for keeping alternative fuel costs low. The loss of availability of a high-grade energy resource then causes higher alternative fuel costs. The use of high-grade energy resources for keeping the costs of alternative fuels low can be called an entropy subsidy.

In 1971 a Bureau of Mines Technical Report, Katell et. al. (1971), showed in an in-depth economic and technical analysis that it was feasible to produce oil from oil shale at $3.74 per barrel. A little more than ten years latter, in 1983, after the oil price shocks, an oil shale project consortium projected that it could produce oil from oil shale at no less than $60 per barrel Graham (1983). The second price estimate using a price index represents a real increase of about 6 times the original estimate, i.e. a 6 fold increase in the real cost of oil from oil shale. The change in price estimates represents a very unexpected change in costs.

---

* Reprinted from *Energy Policy*, Volume 26, Number 2, Douglas B. Reynolds, "Entropy Subsidies," pp. 113-118, Copyright © 1998 with permission from Elsevier Science

I believe that the cause of this incredible price change is the loss of cheap oil supplies that subsidized the cost of the alternative energy resource. As oil supplies tightened, there was less oil available to use in the oil shale process. With less oil, the cost of inputs into producing shale oil, including capital, labor and energy inputs, went up causing oil from oil shale to be more costly than expected. Our conclusion is that the cost of alternative energy resources cannot be easily predicted, when low cost, high quality energy inputs are no longer available.

ENERGY GRADES

Reynolds (1994) defines a physical concept for energy called "energy grades." Energy grades are actual characteristics of energy. Those characteristics are in turn inputs into production processes or inputs into technological mechanisms. Using the concept of energy grades, it can be seen that an energy resource actually provides two inputs into production: energy content (i.e. British Thermal Units [BTU's]) and the characteristic of the energy (i.e. energy grades). Therefore whenever we use technology, we must use energy in order to run it, but the characteristic of that energy is important in determining how easily the technology can use the energy.

Energy grades are categorized into four types: "Area grade", "Weight grade", "Volume grade", and "State grade." These grades define characteristics of energy that determine how well energy can be used by different technologies. Consider first the area grade. The area grade is defined as BTU's per acre of the energy resource in its original state. For example there are some Texas oil fields that have over 500 billion BTUs of oil in a single acre of land underground. On the other hand, many forests have only about one to five billion BTUs of energy per acre. Therefore forests have a lower area grade. The more BTU's per acre of the energy in its original state, the higher is the area grade. The reason why it is important to have a high area grade is because it costs less to obtain energy the higher the area grade is. For example, it can cost as little as one cent per 100,000 BTU's or even less to bring oil to market but wood costs 10 cents per 100,000

BTU's or more because of the lower area grade. It takes more capital and labor to get the lower area grade wood to market then the higher area grade oil.

The next energy grade is the weight grade. The weight grade is defined as BTU's per pound of energy resource. For example a pound of oil has about 20,000 BTU's, but a pound of dry wood only has about 8,000 BTU's. Oil then has a higher weight grade than wood. The higher the energy content per pound of energy, the higher is the weight grade. The weight grade is important because it determines the cost of using energy. Often times a piece of technology such as an automobile or an airplane must carry its energy source with itself. When technology can use a high weight grade energy source, it can usually perform better by either lasting longer or having less weight to carry.

The volume grade is defined as BTU's per cubic feet of volume. We measure the volume grade at standard atmospheric pressure and temperature. For example a cubic foot of oil at standard temperature and pressure has about 1 million BTU's, but a cubic foot of wood has only about 250,000 BTU's. Another example is natural gas. At standard atmospheric temperature and pressure, natural gas is extremely difficult to carry for an automobile. But even when the gas is compressed to 3000 pounds per square inch, it still only has about 177,000 BTU's per cubic foot compared to oils 1 million BTU's per cubic foot. This is why propane automobiles are less convenient than gasoline automobiles. The more BTU's per cubic foot of volume for a given energy resource, the higher is the volume grade. The higher is the volume grade, the more service any piece of technology can extract from each BTU of energy input.

Finally, consider the state grade. The state grade is defined as the physical state of the energy resource at standard atmospheric pressure and temperature. Standard pressure and temperature is used because that is the usual temperature and pressure where society is using the energy resource. An energy resource can be in four states starting with the highest state, which is the liquid state. For example oil is in a liquid state at standard atmospheric pressure and temperature. Lower states in descending order are the gas state, like natural gas, the solid state, like coal, and finally the field state. The field state has to do with energy in the form of fields such as electric fields for electric energy or radiation fields for solar energy. The field state grade is considered the lowest grade because of the difficulty in storing such an energy resource in that form. The higher the state grade, the more service or output that a technological mechanism can produce. The lower the state grade the lower the output possible.

An example of the difference in state grade is like a solar driven car versus an oil fuel driven car. The solar car can only work when the field energy is in operation, i.e. when the sun shines. It is difficult to store the field energy. It might be possible to produce electricity with solar power and then store the electric energy in batteries, however batteries for an electric car are heavy, and they severely limit such vehicles especially on hills or rough terrain. A gasoline driven car though depends on a liquid fuel that is lightweight, easy to store and easy to use. It is easy to see that an oil-fueled automobile is vastly superior to a solar powered automobile for convenience and costs. Certainly solar and electric vehicles are feasible, but oil fueled cars are superior in economic and technological performance.

Note also that electricity as a type of energy is very useful in such applications as telephones and TV sets. It can be argued that electricity especially in these applications is more valuable than a liquid fuel, and hence it is a higher-grade. I define here the state grade of an energy resource in its source state. Electricity does not occur naturally but is a convenient transformation state of energy. All energy resources can produce electricity, but not all energy resources can be used by an airplane. Therefore I believe that the field grade is the lowest state grade.

To see an example of how higher-grade energy resources can produce higher outputs or better service consider when England went from using wood to using coal for steel production in the 18th century. Coal is a higher-grade energy than wood. The higher area grade of coal allowed steel producers to use larger scale iron production capital at that time and obtain increasing returns to scale. In other words, bigger iron works were possible with coal because the iron production facilities can stay in one place and not have to constantly move to new areas of energy production. Bigger iron production mills can produce iron more cheaply. With cheaper iron, the economy could expand more, people could buy more iron products, and so on. In general then in every energy grade category, the higher the energy grade, the better the technology can perform that uses that energy. Table 6.1 and Figures 6.1, 6.2, 6.3 and 6.4 from the last chapter show how different energy resources compare to each other. The higher the energy grade, the more efficiently any given technology can run, and the more cheaply it costs to run.

ENTROPY AND ENTROPY SUBSIDIES

I further consider the concept of energy grades in terms of entropy. Entropy is usually measured by using temperature differentials in order to define how much useful work that can be extracted from a system, i.e. how much work output is available. Another way to view entropy is simply as the amount of useful work or output that a technological system can extract from its energy inputs. The higher the entropy level of energy, the less output it can produce. Energy that is high-grade has low entropy, hence it can produce more potential output per BTU of input. Conversely energy that is low-grade has high entropy, hence it can produce less potential output per BTU of input. In other words, we can think of energy grades as an entropy characteristic.

Next the term entropy subsidy is defined. Consider for example international trade. When one country subsidizes one of its industries such as the farm industry, that country pays money to that industry so that the industry can sell its products at a lower price and at a relatively higher output. Other countries consider such money payments for production as a subsidy of production. In a similar manner, when an economy goes from a low-grade energy input to a high-grade energy input, the costs of production in general become lower. The lower cost production is a result of the better physical characteristics of the higher-grade energy input. High-grade energy inputs will create more output at lower costs.

Since low-grade energy inputs create higher costs and high-grade energy inputs create lower costs, then we can say that the high-grade energy input subsidizes production costs. It is like having a government give you money to produce. Similarly, it is like replacing a poor worker with a more valuable worker who, because he is stronger or smarter, can produce more output than the lesser worker. Since the grade of energy determines its level of entropy, we can call this subsidy an entropy subsidy. Note that such an entropy subsidy does not only cut costs in one industry, but in all industries. An entire economy therefore benefits from an entropy subsidy. We can consider this entropy subsidy as a part of an economic value theory of energy.

THE EXAMPLE OF OIL SHALE

Consider the example of oil shale in order to see how the phenomenon of entropy subsidies works. An initial analyses of the increase in the price of shale oil might identify the cause as being because new technology was not very well developed. Subsequently, with new oil shale technology, not unlike with new military weapons systems, the cost of shale oil would naturally rise due to design changes. However, the Bureau of mines had already experimented with producing oil from oil shale in the 1940's and 1950's, Duncan (1965). What is more, Germany during W.W.II and South Africa in the early 1960's had been able to transform coal into oil, which involves similar processes as oil shale conversion. Furthermore, mining techniques and partial refining needed in the oil shale processes were very well understood for many decades prior to the 1971 report. Certainly new technology could make converting oil shale into oil cheaper, but it is hard to believe it could make it more costly with such a mature technology.

Another possible cause of the price increase is that land and water scarcity created higher than expected costs. If land and water were not plentiful, they would be more costly causing the costs of shale oil to rise. Here again however, there could not have been such scarcity since both the 1971 and the 1983 projects were costs for a single operation. Even though land, water and mineral rights can become scarce, there was certainly adequate quantities of these inputs for a single operation. Therefore, scarcity of land and water cannot be a factor in the stated rise in costs. There has to be another reason, and that is the loss in entropy subsidies.

The US economy as a whole in the 1970's saw higher prices due to the shortage of the high-grade energy resource of oil. As the availability of the high-grade energy resource of oil declined, there was an entropy subsidy loss to the entire economy causing prices to rise. As prices rose through out the economy, the costs for shale oil rose too. This is because the high-grade energy resource of oil was subsidizing the costs of the low-grade energy resource of oil shale. When there was less of the high-grade energy resource of oil, the shale oil cost more. The reason for such large increases in prices is that the process to create fuel from oil shale is 12 times as labor to energy intensive and 3 times as capital to energy intensive as creating fuel from oil, Rocks (1980). The higher capital and labor needs of converting oil shale reflect the lower-grade of that energy in comparison to oil. However, these inputs themselves depend on energy. So if oil was subsidizing shale oil input costs, the less oil there was, the less subsidy there was and the higher the input costs for shale oil were. Then the higher input costs caused the higher price for shale oil.

Indeed every aspect of producing shale oil has direct and indirect energy inputs, especially oil inputs, so that as the price of oil goes up, the price of shale oil will go up. For example the real price of labor is what labor can buy with its wage. When labor buys food, that also requires fuel inputs. Food production uses farming equipment that uses oil based fuels. As the price of oil goes up so does the cost of producing food, making real food costs go up. If labor is to have constant buying power, then their real wages must increase to pay for the real increase in food prices. This means that as oil prices increase, the cost of labor increases making oil shale processing more costly. Similar arguments can be made for capital, material and even energy inputs into the shale oil process.

In other words, estimating the final cost of shale oil is rather like how communist governments tried to plan their economies. They would decide the demand for goods and then try to plan out all the inputs, including inputs that depended on other output goods, in order to produce the demand. Such planning was so complex and indeterminable due to the fact that many outputs were needed as inputs, that if it was not for black markets, many communist governments may not have survived as long as they did. The same kind of inefficiency is inherent in estimating the price for alternative energy resources. As you use less of a high-energy resource for inputs into producing a low-grade energy resource, then you lose the entropy subsidy of those inputs. The energy inputs start costing more causing the output to cost more. The higher cost energy output is then used for its own inputs creating a cycle of cost increases. Since there is a very complex web of energy inputs used for the final energy output, including energy inputs for labor and capital and so on, then the loss of the entropy subsidy is very pervasive. This is the cause of such a large increase in the price of shale oil.

The entropy subsidy loss problem will occur for almost any substitution of a high entropy energy resource for a low entropy energy resource in the same manner as it occurs with oil shale substitution for oil. Alternatively, a transition from a low-grade energy input to a high-grade energy input gives an economy tremendous punch such as England experienced during its switch from wood to coal in the 1700s.

I also believe that technology will help with entropy subsidy losses. For example after 1983, oil prices declined rapidly due to a combination of factors. One main factor was technological progress, which created better oil conservation technology and new oil production techniques to find and extract more oil. This same kind of technological progress will certainly also create lower costs for shale oil and other energy alternatives. However, given only the limited technological progress from 1971 to 1983, it is clear that when technology does not have time to progress, that there is an entropy subsidy loss phenomenon when industry replaces a high-grade energy resource with a low-grade energy resource. Technology may be able to over come a loss in entropy subsidies, yet the phenomenon of entropy subsidies for a given set of technologies exists.

Another problem with losing high-grade energy resources are environmental costs. Often a high-grade energy resource such as oil or natural gas is cleaner to use than a lower-grade energy resource such as coal. It is difficult to define an environmental grade for energy, but environmental problems are also a cost of using energy. Therefore, as the economy transitions from high-grade energy resources to lower-grade resources, we will have to incur the additional cost to the economy of higher pollution costs. If high-grade energy resources are cleaner to use, then that is another subsidy for the economy that can be lost during an energy transition.

CONCLUSION

It is possible that technology can compensate for the loss of entropy subsidies in the economy as we transition from high-grade energy resources to low-grade energy resource. We still must realize however that the entropy subsidy exists as a phenomenon in nature. Therefore, until we actually start producing substantial amounts of low-grade energy resources such as shale oil or heavy oil and use them throughout the economy, we simply cannot know their true cost. We cannot know how much technology will be able to compensate for the entropy loss. The entropy subsidy may cause energy prices to rise higher than expected. The coming transition from oil to oil alternatives may be rocky because of the entropy subsidy loss. We should at least be aware of the possibility of a difficult energy transition, especially since the entropy subsidy loss looked to be so big in the 70's and early 80's with shale oil and other alternative energy resources.

REFERENCES

Duncan, D.C. and V.E. Suranson (1965). US Geological Survey Circular no. 523.

Graham, Sandy (1983). "US Pumps $2 Billion into States Oil Shale" Rocky Mountain News. Denver, July 31, pp. 1, 22, 25.

Hotelling, Harold. (1931) "The Economics of Exhaustible Resources," Journal of Political Economy, Volume 39, Number 2, April, 1931. pp. 137 - 175.

Katell, Sidney and Paul Wellman (1971). Mining And Conversion of Oil Shale in a Gas Combustion Retort, Bureau of Mines Oil Shale Program Technical Progress Report - 44, October, US Department of Interior.

Reynolds, Douglas B. (1995). "Energy Grades and Economic Growth." The Journal of Energy and Development, Volume 19, no. 1 or 2.

Rocks, Lawrence (1980). Fuels for Tomorrow, p. 27, Penn Well Publishing Co.

Solow, Robert M. (1974). "The Economics of Resources or The Resources of Economics," The American Economic Review, Vol. 64 no. 2. American Economic Association, Princeton, May, pp 1 - 14.

Chapter 8: The Energy Utilization Chain: Determining Viable Oil Alternative Technology*

ABSTRACT

This article examines the continuing debate over the limits to economic growth focusing on oil, oil technologies, and the services provided by those technologies. Since there is a growing consensus that oil production will be in short supply in the next 30 or so years, then it is important to examine all viable oil alternatives. I use the concept of the Energy Utilization Chain (EUC) to evaluate oil alternatives and their effect on the economy. The EUC effectively makes the connection between a resource shortage and economic growth. I conclude that all oil alternatives now and in the next 30 to 40 years will create a declining economy.

INTRODUCTION

It is now 25 years since the Arab oil embargo, during which oil prices quadrupled within one year. Although the causes behind the price increases are complex, still a potential future oil shortage is possible. If there is another oil shortage, I examine how exactly technology can over come it. Mikesell (1995) says, “The debate over the limits to growth has been rendered meaningless by the failure… to define growth, … to specify the components essential to growth… and to analyze the relationship between growth … and natural resources.” In this paper, I will attempt to define at least one aspect of growth, to specify at least some of the essential inputs to growth, and to analyze the relationship between growth and natural resources. Out of all the resources that produce economic growth, the most important are energy resources. Energy determines growth. The analysis here will use a concept called the Energy Utilization Chain (EUC), which explains how energy provides energy services to society, creating growth.

With less oil in the future, there will be potentially less energy services available and the economy could decline. There are possible technologies for reversing such decline should an oil shortage occur, however after more than 100 years of using oil, technological advances has failed to find a viable oil alternative with a lower cost than oil. Technology is better. Yet society is as dependent on oil as ever. If oil prices were higher today, than maybe there would be more incentive to develop and use alternatives to oil. Never the less, with a given price of oil that has been constant in real terms for about 100 years, technology has not found any good alternatives for oil.

Tilton (1996) says there are 2 paradigms for the limits to growth debate. The first paradigm is the fixed stock paradigm where mostly physicists claim that as we run out of resources, we will have economic decline. The second paradigm is the opportunity cost paradigm where mostly economists claim that as one resource depletes, other optional resources and better technology emerge to replace it, all be it at a slightly higher cost. The general idea of the second paradigm, that opportunity cost determines scarcity, is true however it is the physical potential for technology to be able to reduce opportunity costs for oil or oil alternatives that is limited.

THE ENERGY UTILIZATION CHAIN (EUC)

The way society uses energy is called the Energy Utilization Chain (EUC). Each link in the EUC works together to make up or build the economy and create growth. The links are as follows:

Link 1. Energy Source. This link of the EUC identifies the energy source. In other words society must be able to identify an energy source and have access to it before society can use that energy.

Link 2. Energy Conversion. This link of the EUC is the conversion of energy. Here is where the economy converts or transforms an energy resource into a more useful form.

Link 3. Energy Consumption. This link is where the economy consumes energy. Society must consume energy in order to create goods and services. Note that conservation of energy resources is also a part of consuming those resources. Conservation allows us to consume less. The less we need to consume, the more that is available for other uses.

Link 4. Energy Service. Finally, the ultimate end of using and consuming any energy resource is to provide a service to society. For example, using gasoline to drive a car provides transportation services. In general, the more services there are from energy, the more goods and services the economy can provide as defined by the GDP and the greater is economic growth. The most important energy service as far as oil is concerned is transportation services and providing Large Independent Mobile Machinery (LIMM) operations. So I will concentrate on those services.

Transportation and LIMM services are crucial for economic growth. Almost all aspects of our economy require or provide transportation and LIMM services. Indeed throughout recorded history, they have been needed. From the boats on the Nile used to carry stones for the great Pyramids of Egypt, to Roman centurions and their horse carts, to modern Earth moving equipment, the level of economic accomplishment is almost always been defined by transportation and

LIMM services. If transportation and LIMM services are worse or more costly then that implies a decline in GDP.

Oil more than any other energy source is the most perfect energy for transportation and LIMM services. A reduction in oil use by definition reduces transportation and LIMM services and must cause a recession unless there is an alternative. Alternatives can be defined by using the EUC concept. However, all conceivable oil alternative EUC's are either not viable or will create economic decline in order to be used.

TECHNOLOGY TYPES

In order to explain how technology can reverse the effects of using less oil, energy technology types are defined. These technologies in terms of their purpose, closely parallel the EUC. These definitions can help distinguish which new technologies will more directly alleviate an oil shortage and which will not.

Link 1 Technologies.

These are technologies for getting new energy resources. They include the following.

Research Technology: This includes how to discover new energy resources or how to get or to use existing energy resources such as possibly new nuclear fuel resources like those used in nuclear fusion. Already, science has more or less charted all resource combinations on the Earth and solar system.

Exploration Technology: The next technology type is exploration technology, which is technology used to find resources. Exploration technology can be as sophisticated as satellites using infrared cameras to look for minerals in the ground, or as simple as people drilling for oil prospects.

Extraction Technology: The next technology type is energy resource extraction technology. This is the technology firms use to extract minerals or other resources. This includes such technologies as enhanced oil production technologies, or even new electricity production technologies.

Link 2 Technologies.

These are technologies for energy conversion. Energy conversion technology converts or transforms energy resources. While converting oil into gasoline has been a very important conversion technology for our past, new conversion technologies might include transforming coal into gasoline.

Link 3 Technologies.

These are technologies for consuming energy. They are defined as technologies that provide a service to society while they consume resources. One example of this is a coal or electric fueled car.

Conservation Technology: Conservation technology is really a part of consumption technology where energy is consumed more efficiently.

THE 5 MAJOR EUC OPTIONS FOR OIL

In order to replace the oil EUC where oil is the primary energy resource that produces transportation and LIMM services, alternative EUC's that can produce transportation or LIMM services are needed. Basically there are five viable EUC's that can replace the oil EUC. These EUC's are the following:

1.) The Oil Use EUC. This EUC concentrates on link 1 of the EUC by simply finding and extracting more oil without regard to oil substitutes.

2.) The Oil Substitutes EUC. This EUC concentrates on step 1 and 2 of the oil EUC. This is where oil is substituted for directly by using fuels from such energy sources as coal and oil shale and natural gas.

3.) The Oil Conservation EUC. This EUC concentrates on step 3 of the oil EUC. This is where less oil is used by for example having higher mileage cars.

4.) The Electricity EUC. This EUC concentrates on steps 1, 2 and 3 of the oil EUC. This is where oil is replaced completely by using electricity and even solar power.

5.) The Life Styles EUC. This EUC concentrates on step 4 of the oil EUC. This is where less oil is used by changing our life styles. Usually most life style changes require a lower level of GDP production and so by definition means a decline in the economy.

OIL USE EUC

The main idea of the oil use EUC is to simply find and extract more oil. Currently the world is estimated to have about 2.4 Trillion Barrels of ultimately recoverable oil reserves. However, more oil can be found and produced with better technology. As far as new exploration and production technologies for oil are concerned, there does not look to be any great advances. Already US oil production has declined for 30 years even though all kinds of new exploration and extraction technologies have come on line during that time and even though Alaska Prudhoe Bay Oil came on line during that time.

On the other hand, if prices rise in the future, more oil offshore or in the arctic may be found where it is expensive to search for and produce oil. However, given a fairly high price of oil, technology has not been able to over come declining oil production for 30 years in the US. If 30 years of technology change has not been able to increase oil production in the US, then it probably has not substantially increased potential production in the rest of the world.

If technology cannot increase potential world oil production for 30 or 100 years greatly above what is technologically feasible now, then we must expect some sort of oil shortage. Given the predicted ultimately recoverable world oil reserves, this shortage will occur within 100 and more probably 30 years. That means higher oil prices and less supply, which by necessity would reduce transportation and LIMM services causing decline in world GDP. Therefore this EUC does not look to be viable. One other possibility though is to produce more oil from oil shale, heavy oil, and tar sands. However, oil from these sources is not crude oil. These are substitute energy resources, which we will look at next in the oil substitutes EUC.

OIL SUBSTITUTES EUC

The oil substitutes EUC depends on changing steps 1 and 2 of the oil EUC by substituting alternative liquid, gas and even solid fuels directly for oil in the oil EUC. In this way, steps 3 and 4 of the oil EUC require few if any changes. The economy merely switches fuels. The main direct fuels available for this substitution are: Natural Gas, Methanol, Synthetic Fuels (Synfuels), and Ethanol. I examine these fuels and the technologies available for using them.

Reynolds (1994) explains how energy resources have different grades that affect how well they perform in an economy. Low-grade energy resources are less useful to an economy than higher-grade resources. There are four grade types: 1) The weight grade (British Thermal Units (BTU's) per pound of the energy resource). 2) The volume grade (BTU's per cubic foot of the energy resource). 3) The area grade (BTU's per acre where the energy resource is found in its original state). And 4) the state grade (the original physical state of the energy resource as a liquid, gas, solid or energy field). These grades are fundamental in determining how well alternative energy resources can substitute for oil.

NATURAL GAS

Natural gas has some relatively good energy characteristics. It is a gas, so it has a fairly high state grade. It also has good weight and area grades. However, its volume grade is very low. For example, oil has a volume grade of 1 million BTU's per cubic foot where as gas, even when compressed to 3000 psi only has about 0.2 million BTU's per cubic foot. This makes gas difficult to carry. Therefore, most gas vehicles have a low mileage range or a huge bulky tank that they must carry around. So far, natural gas has not been a convenient transportation or LIMM fuel.

METHANOL AND GAS TO LIQUIDS

Since firms can convert natural gas into methanol or another fuel called white gas, using the Gas To Liquids (GTL) technology, and then transport the methanol or GTL relatively cheaply, then it would make sense to use methanol or GTL directly to replace oil rather than natural gas. Considering methanol, it has all around good characteristic grades. It is a liquid. It has a high area grade since it comes from natural gas. However, its volume grade is somewhat low at .43 million BTU's per cubic foot compared to 1 million BTU's per cubic foot for oil and gasoline, which decreases its usefulness. GTL is about the same. Currently, methanol or GTL costs about $50 per Barrel of Oil Equivalent (BOE) to produce which could be competitive should oil prices go high again, but there are other problems.

One problem with methanol and GTL is that there are losses of converting natural gas, which run at about 40%, i.e. only 60% of the natural gas energy is converted. These losses can add to costs if gas producers must pay for exploring for greater reserves of gas. Also since methanol and GTL are made by a complex conversion process, then as energy prices in general go up, so will inputs into the conversion process which will in turn cause the price of methanol and GTL to increase. Reynolds (1998) calls this inflation cost spiral for energy the entropy subsidy problem. All energy conversion processes have this problem. When the cost of oil goes up it makes all other goods and services in the economy more costly. When these more costly goods and services are used as inputs for energy conversion, then the new fuel that is produced is more costly then expected and that causes higher costs for inputs and so on.

SYNFUELS

Synfuels is short for synthetic fuels. These are liquid fuels that firms convert from solid fuels such as coal, oil shale and tar sands. Industry can convert these low state grade energy resources into higher state grade resources, so that they could replace oil. However, since they start out as low state grades, they require expensive transformations before firms can convert them into a liquid or a gas that could be used in the oil substitutes EUC. Usually, oil from synfuels is 3 times as labor to energy intensive and 10 times as capital to energy intensive as crude oil (Rocks, 1980). Again we get the entropy subsidy problem where when fuel prices go up, then so do labor and material prices. These input prices in turn cause the labor and material intensive synfuels to cost more than originally expected. Thus, any attempt to estimate the cost of synfuels is automatically an incorrect cost because it does not nor cannot account for the entropy subsidy problem.

SHALE OIL: According to estimates, the US has about 3 trillion barrels of known recoverable reserves of shale oil. The rest of the world has about 2 trillion barrels. However, most shale oil production requires large mining or strip mining type operations. In order to produce millions of barrels of oil per day, which is what the US will need, industry will have to build several hundred such large mining operations. Most shale oil production will occur in a place called the Piceance creek basin. In this area in Western Colorado, the richest US oil shale reserves exist which exceed 200 billion barrels and which would probably produce 70% or more of total US shale oil production. See Cuff (1986). However, the Piceance creek basin is only 1,700 square miles. That means that an area the size of about Delaware, but with huge cathedral like mountains and dry barren country, would have to have thousands of huge mining and milling operations each operating on hundreds or thousands of acres. If the US were to depend on shale oil as a major source of energy, then this area would become one giant industrial wasteland.

HEAVY OIL: It is often suggested that heavy oil should be lumped together with conventional crude oil to determine oil reserves. This is a mistake. Heavy oil is a substitute for conventional crude oil and it is nothing like crude oil. Heavy oil is crude oil that is thicker and heavier than conventional crude oil. In fact these oils are so thick that they must be considered as a solid-state grade and not a liquid state grade. Because they are a solid, they require very high temperatures before a pump can pump them out of the ground. In other words, heavy oil has to be converted from a solid to a liquid just to get it out of the ground. Usually firms set the oil on fire to do this. That is, some of the underground oil is ignited, often by pumping in oxygen, which allows the remaining oil to become less viscous and flow out of a pump. Currently about half of the world's known reserves of heavy oil, about 300 Billion barrels, reside in

Venezuela and another 30% in the Middle East. So Venezuela has about ten times the proven reserves of conventional oil in the US.

However, Venezuela has all its reserves located in a small area that is less than 500 miles long and 45 miles wide in the so-called Orina River belt. Furthermore in order to produce heavy oil, a firm can only extract 5 to 100 barrels of oil a day from a single producing well, in contrast to being able to produce up to 10,000 barrels a day from a conventional oil well. This makes heavy oil costly. But the real problem is that since Venezuela's heavy oil belt is so small geographically, then many of these heavy oil wells will be located very close together in order to produce substantial amounts of the oil. This brings up the same problem that oil shale has which is that a lot of capital equipment has to be located close together so that there is not enough room to work in. That increases costs for firms.

TAR SANDS: Tar sands are another interesting synfuel concept. Canada has the potential to produce almost 1 trillion barrels of oil from tar sands. Venezuela may have 2 trillion barrels. The US has about 30 billion barrels of reserves and Russia's reserves are unknown. So far, engineers have not been able to produce commercially viable quantities of Tar Sands. It is expected though that Tar Sands potential depends on converting the tar sand into a liquid in place under the ground. Such a process is not yet feasible. It would be impossible to tell the final cost.

COAL CONVERSION: Fuels from coal can be made by either gasifying the coal, called gasification, or by liquefying the coal, called liquefaction. In these methods, the final product is often further converted or refined into higher-grade gases and liquid fuels. Some US companies feel like they can use these technologies to produce oil at about $50 per BOE depending on how much government support they get. Again there will be an entropy subsidy problem so that once oil does become $50 per BOE, the cost of coal conversion will become higher.

Another option often mentioned for coal is to use coal directly as a fuel source. In this way, we could have cars that run on coal directly just like steam locomotives used to. However, coal is a solid. Because of this, it is almost impossible to use coal directly in an internal combustion engine such as most cars and diesel trucks have. The only option for using coal then is for an external combustion engine. However such engines are heavy. They are in fact so heavy that a lot of the fuel is used just to carry the engine and the performance of such vehicles would be very poor. Many other alternative vehicles, such as electric vehicles, discussed later, are better than coal vehicles for most transportation needs. Thus coal or any other solid fuel is a poor fuel option for the oil substitutes EUC.

BIOMASS

Biomass is technically the dry weight of living matter. For energy uses it includes such energy sources as trees or plants in forests or plantations; crops or crop residuals; aquatic plants or algae; and animal and human waste. Firms can use these to produce fuels like methane, methanol and ethanol, as well as other less common fuels. However, if the US used a lot of its farmland, or even a small percentage of it, for the production of energy, that would lead to higher food prices or even food shortages around the world, as much of the world depends on US food production. An even bigger problem, though, is the environment. Many biomass energy production schemes require huge land area changes and water right diversions to produce relatively small amounts of energy. Soil erosion, water depletion, and changes in global weather patterns could be a problem.

Biomass sources such as trees and crops have a very low area grade and, until they are converted into a fuel, a low weight, volume and state grade as well. In addition, the time of availability for most biomass is either in the summer or the fall. Biomass conversion plant must either stand idle much of the year or have extra biomass stored in reserve. However, biomass decomposes rapidly causing huge losses and making a plant less economical to run. Plus the quantity of biomass from any given source may be very low, like for example using biomass from animals that are on a small farm, causing a biomass conversion plant to have low economies of scale.

The oil substitutes EUC options are all more costly than oil and potentially by several times the price of oil. They all have worse energy grade characteristics so that no matter how good technology becomes, they can never compete with oil. Looking at technological change over the last 30 years, we see that oil substitute EUC technologies have not progressed very rapidly and certainly do not look to be on the verge of any major break through for the next 30 to 100 years.

OIL CONSERVATION EUC

The Next EUC is the oil conservation EUC. In this EUC, the economy tries to conserve oil, or even alternative fuels, by using better conservation techniques. This creates two effects. First, by conserving oil there is less demand pressure on world oil supplies, which tends to lessen an oil shortage and so lower oil prices. Second, when the economy uses less oil, it then needs a lower percentage of its GDP to purchase oil even when oil prices are high, or in other words, it helps producers and consumers save money on their oil bill. That makes the economy stay strong.

Already today there are direct injection diesel engines that get upwards of 100 mpg (Goldemberg, 1987). We can call these 100-mpg cars. So, if the economy could use more of these kinds of 100-mpg type cars and even find other more oil efficient technologies, then maybe we could put a dent in the oil crisis. It may be possible to have even more efficient cars such as 200 or 300-mpg cars. However, the efficiency of a motor depends on how cold the intake air is and how hot the fuel becomes when it combusts. Since we know what the temperature is of the intake air, somewhere around room temperature, and since we know how hot gasoline is when it combusts, then we know the highest efficiency that a motor can ever attain. For the most engines that efficiency is 40%. This puts a limit on mileage that is possible.

In general, though oil conservation technology such as the 100 mpg car create a loss in transportation services since they are not nearly as comfortable or as safe to use as regular automobiles. Also LIMM services are lower when machines are required to be much more fuel-efficient. Another oil conservation method would be to use more mass transit, which is really a life style change, which I consider last.

ELECTRICITY EUC

Another option for replacing the oil EUC is to change step 3 entirely, that is change the type of fuel. One alternative fuel type would be electricity. Electricity is a field type energy state grade, which is the lowest state grade. It is difficult to store and carry a field type energy, and since the method of transportation works best when a transportation device carries its own energy source with it then that makes electricity hard to use.

One of the biggest concerns for an oil shortage will be finding ways to run LIMM's such as farming tractors, construction machinery, and mining machinery. These oil uses are crucial. If we cannot use such machinery, then food prices, building costs and mineral costs would increase and the economy could go into a recession. Besides this, transportation will also be a problem. Electric track vehicles such as electric trains may be one option, but there will still be a need for independent trucks and transportation vehicles. For example, electric trains could haul many goods long distances, but for short distance hauling needs, such as from a train station to a supply warehouse, trucks are almost essential. Without trucks the economy would suffer.

One possible electric technology for transportation is independent electric vehicles using batteries. The biggest problem is the weight of these batteries. For example, one ton of lead batteries or 650 lbs. of zinc nickel batteries is equivalent to 1 gallon of gasoline. With such huge weight to energy ratios, much of the energy that is available with electric batteries is used just to carry the energy itself. This especially causes problems when going up hills. While electric vehicles can reduce the need for oil, they cannot totally replace cars and LIMM's. Another option for electric cars is to have them powered by both a battery and a liquid fuel. In this way, when the battery runs low, the liquid fuel such as methanol can take over. This increases the mileage of the vehicle. While this can help make electric cars more useful, it is still an indirect way to burn a liquid fuel. It wouldn't be any cheaper then using the liquid fuel directly. It may conserve the use of oil or other fuel, but it does so at a lower performance. There is a loss of service.

Solar energy along with such energy sources as hydropower, wave power, tidal power, wind power, geothermal energy and all types of nuclear power, as well as generation technologies like fuel cells, thermionic converters, thermoelectric generators, and magneto hydrodynamic generators, are all ways to produce electricity. While many of these energy sources may produce cleaner power, they are never the less only ways to produce electricity. We can already produce cheap electric power using coal and natural gas. It is using electricity to replace oil that is the problem. Even solar cars are essentially useless at night. In general the electricity EUC does not produce very useful transportation and LIMM services. It does not adequately replace oil.

LIFESTYLES EUC

The fifth EUC, lifestyles, calls for changing not only energy sources that we use but also energy services. In other words, this EUC changes all four steps of the oil EUC in order to conserve oil. This means a lower standard of living. The entire economy will have to restructure its way of doing business so that people will have less, earn less, and pay more for goods and services.

As oil prices get higher and higher, it will affect the cost of acquiring raw materials. The price of transporting these materials will increase causing higher raw material costs. This may create more regional markets for raw materials. People will start to depend on local raw materials rather than far away materials. Local small factories will take over large-scale centralized factories so that people will get more of their needs met locally. But while this sounds very ecological and more humane, it really means that we will lose economies of scale and have less income and a lower standard of living.

Many mining and forestry operations use substantial amounts of capital equipment such as cranes, dump trucks, earth moving tractors and so on. Most of these machines use diesel fuel. They cannot adequately use electricity or even other fuels to run on because those fuels are either too heavy or too low of a grade. This means that as the cost of oil goes up, the cost of acquiring minerals from the ground or wood products from forests will become high.

High oil prices will also severely affect the construction industry, which is one of the biggest industries in the US after the auto industry. Much of the large machinery necessary for building projects requires diesel fuel. When oil prices go up this will cause all new construction to be more expensive. That will create higher building costs. This in turn will mean less housing for people as they will not be able to afford as much as well as less new factories and new businesses as these will also cost more.

Probably the most tragic aspect of high oil prices is how they will affect our food source. Most farming and fishing operations in the US use large amounts of oil fuels so that as the price of oil goes up, the price of food will go up too. Third world countries that import much of their food will not be able to afford as much food. They could face food shortages or starvation. Even third world countries that do not import food will still be affected by high oil prices as there is a tendency in the world to use fertilizers from oil in order to grow higher yield crops. One way to counter act high oil prices at least in the US is to use more labor to produce food instead of machinery. But, this will still cost money. Food prices will still be higher and the GDP lower as the economy forces consumers to spend more of their money on food and housing and less on other goods and services.

The single biggest use of petroleum is for transportation. The way we move people, goods and services is the most important characteristic of our entire economy. Hence, our lifestyles depend on transportation. One way to conserve oil would be to have people use more mass transit rather than personal vehicles for commuting to work and running errands. That would save a lot of oil. Usually using mass transit takes twice as long as, or longer than, driving an independent vehicle for personal transit for a similar distance and mass transit is much less convenient. This is true even when there is a well-developed mass transit system. More mass transit also means living closer to cities, which means more noise, pollution and lack of nearby nature, which means even more hidden costs. So, there would be a large life style change if society used more mass transit and that technically means higher prices for goods and services and a lower GDP.

One last life style change is the idea of people working at home on their computer. This is another possibility and will surely increase in the future. Although it is easy to imagine doing a lot of paper work at home, most work places still rely on meetings and other face-to-face interactions such as simultaneously looking at a plan. After all it has been 40 years since the widespread use of Television and still most people go to classes at school rather than watch classes on television.

TRENDS IN ENERGY TECHNOLOGY DEVELOPMENT

I might finish here by noting that the trends in energy technology development look alarmingly slow. For example, in the race to find nuclear fusion, science has yet to reach the break-even point for a sustained period on a consistent basis where by they can produce just as much energy from a fusion reaction as they use to create it. In fact, it looks like nuclear fusion is always at least another twenty years away from being useable no matter how far along they get. Another interesting trend is automobile technology itself. For the last 80 or so years, automobiles have used basically the same type of internal combustion engine. If that trend continues, then miles per gallon ratings will not significantly rise much higher than the best technology that already exists.

Another example of a trend is with synfuels such as oil from coal and shale oil. These synfuels have consistently been expensive. They keep costing more than oil such that it seems like the cost of synfuels is always 50% more than whatever the current price of oil is (Reynolds 1998). This indicates that the costs of oil substitutes are completely unknown. In fact, we probably know more about the costs for putting an astronaut on mars than we do about how much shale oil costs. Engineers simply cannot predict these costs. And they can never predict them because of the inflationary cycle that keeps pushing costs higher due to ever-higher oil prices.

If we look at the trend in computer technology, we see that the speed of computation has increased by many magnitudes in the last 30 years even while the cost of computing has declined. Yet even though oil prices have stayed steady there is no new technologies to replace it without a sever loss in GDP. We must expect these trends to continue. If one believes that it is unrealistic not to expect technology to get better, then it would also be unrealistic not to look at the trends in technological development that have already occurred. Interestingly, one of the most important trends of all in energy technology is that the world has endured two oil price shocks in the last thirty years, three if you include the fall of the Soviet Union, and still we are as dependent on oil as ever. What these trends suggest is that technology cannot be counted on to save us from an oil shortage.

CONCLUSION

Rosenburg (1973) looks at history and shows that there were many instances of resource shortages that were over come with new technology. It is in fact easy to imagine that since oil is an energy resource then once an oil shortage occurs, there will be many alternative energy technologies and many energy resources that can replace it since so many energy sources and so many energy technologies exist. However it is important not to link oil with all types of energy and it is important not to link all technology with energy technologies. In fact all of Rosenburg's examples of averted resource shortages are examples of economies going from low-grade energy resources to higher-grade energy resources. In our age, we are faced for the first time in the history of man with going from a high-grade energy resource to a lower-grade energy resource.

I believe the discussion over the limits to growth debate can be made more meaningful if it can focus on the most important natural resource and its most important uses rather than broadly generalizing about all natural resources and all technologies. What I have done here is to focus on a particular link between a natural resource and economic growth, which I believe to be the most important links of all. Of all natural resources that exist, energy resources are the most important because they are mostly non-renewable and non-recyclable. Additionally, of all the energy resources we have, oil is the most important because its characteristic grades are generally higher than any other energy resource. Therefore, focusing on oil resources is of paramount importance.

REFERENCES

Barnett, Harold J. and Chandler Morse (1963). Scarcity and Growth: The Economics of Natural Resource Availability, Johns Hopkins University Press for Resources for the Future, Baltimore.

Cuff, David J., and William J. Young (1986). The United States Energy Atlas, 2nd edition. Macmillan Publishing Co..

Goldemburg, Jose (1987). Energy for a Sustainable World, World Resources Institute, Washington D.C..

Meadows, Donella H., Dennis L. Meadows, Jorgan Randers, William W. Behrens III (1972). The Limits to Growth, Universe Books, New York.

Mikesell, Raymond F. (1995). "The Limits to Growth: A Reappraisal," Resources Policy, Volume 21, Number 2, pp. 127 – 131.

Reynolds, Douglas B. (1994). "Energy Grades and Economic Growth," The Journal of Energy and Development, Volume 14, Number 2, Spring, pp. 245 – 264.

Reynolds, Douglas B. (1998). "Entropy Subsidies," Energy Policy, Volume 26, Number 2, February, pp. 113 – 118.

Rocks, Lawrence (1980). Fuels for Tomorrow, Pennwell Publishing Co., Tulsa.

Rosenburg, Nathan (1973). "Innovative Responses to Materials Shortages," The American Economic Review, Volume 64, Number 2. Pp. 1 - 14.

Simon, Julian L. (1996). The Ultimate Resource 2, Princeton University Press, Princeton.

Tilton, John E. (1996). "Exhaustible Resources and Sustainable Development," Resources Policy, Volume 22, Number 1/2, pp. 91 – 97.

Chapter 9: The Economics of Alternative Energy Resources*

ABSTRACT

Oil is one of the world's most widely used energy resource, and may be the most valuable energy resource there is. However since the oil price shocks of the 1970's, and since the rise in environmental and global warming awareness, engineers and scientists have attempted to find energy alternatives in order to become less oil dependent. Most oil alternatives have been unsuccessful. One reason why oil alternatives are not more widely used has to do with economies of scale. Oil alternative energy resources have poor physical characteristics, called energy grades, which creates a loss of economies of scale. As the world transitions away from oil to oil alternatives, the loss of economies of scale inherent in alternative energy resources can cause lower economic growth and may even cause decline. In order for technology to successfully over come the loss of oil's high quality physical characteristics there may need to be extremely large increases in productivity.

INTRODUCTION

This article looks at the role of economies of scale in the use of energy. Simon (1996) has long advocated that the ultimate resource is humankind's capacity for innovation, presumably among other things in order to create higher

* "The Economics of Alternative Energy Resources" by Douglas B. Reynolds, working paper.

productivity. When it comes to energy production and use though, there are certain physical characteristics of energy that better technology and innovation may not be able to overcome. Some energy resources have a much greater potential to produce goods and services especially transportation services and the use of Large Independent Mobile Machinery (LIMM). Goldemberg et. al. (1987) show new technologies that will extend the use of energy by increasing energies productivity. Nevertheless, I will show that the physical characteristics of energy resources do create a huge difference in how useful any energy resource is. In particular, oil is far and away the most productive energy resource there is and its physical characteristics create economies of scale that cannot be overcome with better technology.

In 1913 in Highland Park, Michigan, Henry Ford was building one of the first ever factories with an assembly line method. The significance of this event cannot be underestimated in the history of the industrial revolution. The basic idea behind Ford's novel approach was that instead of having ten factories with say 100 *skilled* technicians each, hand building maybe ten cars a week, now you could have 1 big factory with 1,000 *semi skilled* workers produce 1000 cars a week. The monetary investment needed for ten factories is the same as that needed for one giant factory, but there is a ten-fold increase in output. The cost per car declined but the quality of workmanship actually increased. Indeed, production output of cars decreased from 1 car every 728 minutes to 1 car every 93 minutes with the initial factory and eventually to one car every 24 seconds some time later. See Encarta (1998). That is a time decrease of more than 3 magnitudes. No wonder cars became cheaper even as wages rose.

Ford managed this incredible feat of production efficiency by using his new assembly line technique. On the assembly line each worker could specialize at one job with very little skill and produce more than if he had to do several extremely skillful jobs. The idea behind the assembly line is called economies of scale. Economies of scale occur when the size of the factory doubles but the quantity of output increases by more than double. In other words, doubling the labor and capital inputs increased the output more than 2 fold. Ford was not the first to employ economies of scale to increase production per worker, but his was one of the most innovative approaches even if it was a simple idea. Nevertheless, economies of scale is one of the most pervasive forces in all of economics and has been instrumental in human economic ascent. However, economies of scale is not only a force behind technology it is also a force behind the inputs of production too, most importantly energy inputs.

One of the subtlest ways in which economies of scale work has to do with energy inputs into production. High-grade energy resource inputs actually produce greater output and have more economies of scale per measured amount of energy than low-grade energy inputs. It is the characteristics of energy resources that determine their productivity potential. These characteristics are called energy grades. Reynolds (1994) defines energy grades based on general physical characteristics of energy. High-grade energy resources have more potential usefulness than low-grade energy resources. The energy grade affects the economic potential of using any given energy resource by affecting the potential economies of scale for any energy use.

ENERGY GRADES

Energy actually provides two inputs into any economic process: energy content and energy grade. Energy content is defined by the measurement of British Thermal Units (BTU's), which is the amount of heat producible by energy that can raise one pound of water by one degree Fahrenheit. The metric unit for energy measurement is Joules. Energy grade is measured in one of four ways, the "area grade," the "weight grade," the "volume grade," and the "state grade." The importance of the energy grade has to do with how easily and efficiently the energy content can be produced and used. In general, high-energy grade resources have the potential to create returns to scale where as low energy grade resources lose returns to scale.

Another way to look at the energy grade concept is that every energy resource has energy content measured in BTU's. However, a high-grade energy resource can produce more economic output per BTU than a low-grade energy resource. It is like the difference between many small factories and one giant assembly line factory. With both factory configurations, the amount of manpower and even capital is the same, but the large factory has economies of scale and is able to produce more. Just so two different energy resources can be rationed so that each has the same amount of BTU's, i.e. energy content. However the high-grade energy resource will have a more concentrated energy content, which will allow it to incorporate economies of scale in its use. The high-grade energy resource will be able to produce more economic output per BTU than the low-grade energy resource, because of its higher energy concentration. To see how this works, consider each of the following grades.

The area grade is the amount of BTU's per area of the energy resource in its original state. For example forests generally have 1 - 5 billion BTU's per acre of wood. Coalmines have 10 to 1000 billion BTU's per acre. The more BTU's per acre, the higher the area grade. A higher area grade creates economies of scale in the production of energy. The reason why it is important to have a high area grade is because it costs less to obtain energy the higher the area grade is. For example, it costs about $1.5 per million BTU's to bring coal to market, but wood costs $5 to $20 per million BTU's because of the lower area grade. It takes more capital and labor to get the lower area grade wood energy to market then the higher area grade coal energy.

The weight grade is BTU's per pound of the resource. Coal has 10,000 BTU's per pound while oil has 20,000 BTU's per pound. By having a higher weight grade fuel, a vehicle such as a train can get better performance since it has to carry less weight per BTU. This gives economies of scale for the concentration of energy. A higher concentration of energy allows for better performance of any given energy use. The volume grade is BTU's per cubic foot of the resource. For example, oil has 1 million BTU's per cubic foot while natural gas at standard atmospheric pressure and temperature has 1,000 BTU's per cubic foot. If gas is compressed to 3,000 psi., it has about 177,000 BTU's per cubic foot. The volume grade is often a determining factor in the economies of scale for transporting energy. Oil can be transported 6,000 miles in super tankers for about $0.50 or less per million BTU's while natural gas costs about $3.00 per million BTU's when transported as a super cooled liquid.

The state grade is the physical state of the energy resource at standard atmospheric pressure and temperature. For example at 68 degrees Fahrenheit at sea level, oil is a liquid, methane natural gas is a gas and coal is a solid. Another state is the field state. The field state has to do with energy in the form of fields such as electric fields for electric energy or radiation fields for solar energy. Any energy resource such as solar power, wind power or nuclear power that is in the form of a field gradient is a field state grade. The field state grade is considered the lowest grade because of the difficulty in storing such an energy in that form. The highest state grade is a liquid such as oil. Lower states in descending order are the gas state, like natural gas, the solid state, like coal, and finally the field state. The higher the state grade, the more service or output that a technological mechanism can produce. The lower the state grade the lower the output possible.

An example of the difference in state grade is like a solar driven car versus an oil fuel driven car. The solar car can only work when the field energy is in operation, i.e. when the sun shines. It is difficult to store the field energy. It might be possible to produce electricity with solar power and then store the electric energy in batteries, however batteries for an electric car are heavy, and they severely limit such vehicles especially on hills or rough terrain. A gasoline driven car though depends on a liquid fuel that is lightweight, easy to store and easy to use. That is why an oil-fueled automobile is vastly superior to a solar powered automobile for convenience and costs. Certainly solar and electric vehicles are possible to use, but oil fueled cars are superior in economic and technological performance.

PAST ENERGY RESOURCES

The usual reason given for human kinds economic achievements is that a series of technological innovations created human economic ascent. See Bronowski (1976). However, another aspect of economic growth is the increasing grade of energy resources that were made available with technology. Human economic history can actually be looked at in terms of past energy resources used, using the concept of energy grades. Energy resources had increasing grades. The subtle cause and effect behind our great human economic advance is that by using higher-grade energy resources, we have achieved greater economies of scale and hence more economic production. Consider than an analysis of various energy resources' potential based on economies of scale. The following are energy resources that society used in the past.

WOOD

Wood energy is one of the original energy sources of society. It is a very useful energy resource in that it is easy to find since it is above ground where as most other energy resources are below ground or in the form of fields such as solar energy. However wood has a fairly low area grade and it has a low state grade. Besides its weight and volume grades being low, it is difficult to transport since it is in such a bulky state. Unless it is chopped up, moving wood is extremely cumbersome. Consequently, for much of history, societies would simply burn forest where they stood, use the wood for heat, use the charcoal for smelting iron and move on. This necessitated that iron-smelting operations were mobile and therefore relatively small time operations.

COAL

The first big break for energy use in society came in 1709 when Abraham Darby invented coking methods for coal. Suddenly, it became possible to use coal instead of wood for iron and steal production. By the mid to late 18th century, the steel industry was a major part of the British economy at the point of its industrial revolution. The change over from wood to coal is a change of economies of scale. Wood has a low area grade, 1- 5 billion BTU's per acre while coal has a much higher area grade, 10 - 1,000 billion BTU's per acre. That is an increase in area intensity of about 2 magnitudes.

To understand the significance of this event of changing from wood to coal, imagine trying to set up an iron producing operation. First of all since wood and iron ore may be far away from each other there is a transportation problem, either the iron ore must be taken to the wood or visa versa. However, due to the bulkiness of wood it was easier to take the iron ore to the forest. So firms had to set up a transportation network from ore mines to forests. Usually the forest was burned downed in order to make charcoal for the iron mills. The charcoal was gathered and the iron was produced and then transported to the city.

However, once a forest was consumed, the entire milling operation had to be moved to a new location in a virgin forest. The transportation networks had to be reset and the small factory operation disassembled packed and moved on. In some cases, the factory was simply left behind if it was made of brick and a new factory built at the new location. This constant moving and rebuilding of operations meant that the iron mill owner did not want to invest in large-scale operations since each operation would either be moved or be left behind useless. In a similar manner, the transportation network between iron ore mines and forests and then to the city also had to be flexible and easily changed. Usually, this meant using small horse drawn wagons.

Once coal came on the scene, a single coalmine had the equivalent thermal energy of 100's of forests. Now the iron and steal factories could simply stay in one place without being moved, and since the firms knew that they could use a single mill operation for a long time they could incorporate economies of scale. Now firms could use economies of scale by locating one giant factory right by the coal mine. Or the factory could be by an iron mine with a rail or water connection to the coal. The transportation network itself could now incorporate economies of scale since it too would not be moved around. One railroad or one water way, which could carry as much as a hundred horse drawn carriages between two mines, could be built and not have to be moved ever. If such connecting railroads did have to be moved, then it would not have been worth the investment and they never would have been built. It was the high area grade of coal that made all these huge investments worthwhile.

These economies of scale in iron and steal were a big factor in the industrial revolution in general. It was not just technology that created these economies of scale, but rather the high area grade of coal. With coal there was a centralized fuel source greater than a hundred forests in one location that could be used as an input for steal production and for transportation such as steamships and railroads. However this was not the end of the line for energy improvements.

OIL

Oil has some of the highest energy grades of all energy resources in all categories of energy grade characteristics. It has a high area grade like coal. It has the highest state grade, a liquid. It has a high weight and volume grade. No wonder it was the most useful energy resource ever used by man. In the late 1800's, the US and Western Europe were using coal energy. The technologies of the day were steamships, steam locomotives and steam engines for industry. By changing from coal to oil, industrial production and services to consumers obtained a tremendous cost advantage due to economies of scale. However, this change created an economy of scale in intensity not in size.

In order to use coal, since it is a solid energy resource, it is very difficult to let the fuel flow into a chamber, burn and flow out. Consequently coal is best used in an external combustion engine like for example a steam locomotive. Oil based fuels are liquids and can flow into and out of chambers easily. They can flow in very small measured amounts, be burned and blown out, consequently they allow the use of internal combustion engines. The difference between an external and an internal combustion engine is one of size and weight. Internal combustion allows a smaller lighter engine that is more versatile. Therefore the economy of scale created with oil-based fuels was an economy of scale of greater intensity of power per pound of engine. Internal combustion engines produced more power per pound. This kind of economy of scale meant that transportation could become lighter weight and more versatile. And it did. The explosion of new transportation vehicles, diesel trains and ships, gasoline cars and trucks and even airplanes, was caused by the tremendous increase in the economies of scale of the intensity of energy utilization that in turn created a profound push to the world's economies and all because of oil's vastly superior high-grade energy characteristics.

Oil had a higher weight and volume grade than coal, so that it gave better performance for transportation. No longer did a transportation vehicle have to carry heavy coal with a heavy external combustion engine. Now transportation vehicles could carry a lightweight fuel per BTU with a light internal combustion engine. Again with the advent of oil as a fuel, we see the industrial revolution take off to great heights. It was not just technology that created so much transportation possibilities, but rather technology along with a high-grade energy resource such as oil. Oil fuels create transportation possibilities never before dreamed of such as trucks, cars and airplanes, which in turn created greater industrial specialization and greater economic mobility. In addition, Large Independent Mobile Machinery (LIMM's) such as tractors, mining vehicles, backhoes and the like, created huge productivity increases. It is hard to imagine how the industrial revolution could have ever gotten off the ground with such a profound take off without oil or coal.

NATURAL GAS

The usefulness of natural gas has been more important in terms of pollution than anything else. Coal is usually cheaper than natural gas in most markets as a pure energy source, due to coal's higher volume grade. However, coal creates environmental problems. Gas being in a gas state grade is able to burn more cleanly than coal, and it can achieve higher conversion efficiencies when used to produce electricity. Nevertheless, gas has mostly been a back up to coal use. It has helped fuel the industrial revolution, but it hasn't been a major instigator.

One use for natural gas has been as a fuel in internal combustion engines. Since natural gas is a gas it can be easily injected into an internal combustion engine and achieve the same power to weight performance characteristics that oil fuels achieve. The problem with natural gas is its volume grade. Being a gas, natural gas only has 1,000 Btu's per cubic foot at atmospheric temperature and pressure, and only 177,000 at 3,000 psi. In other words natural gas is not as good as an oil fuel because its low volume grade makes it very difficult to store. The low volume grade gives gas lower returns to scale. An automobile using natural gas needs to fill up more often or have a much bigger tank. In some situations this is not too difficult a problem, but it does create lower performance.

NUCLEAR

Nuclear power has had pollution concerns due to radioactive waste and radiation leaks. Nuclear power is based on using a radioactive fuel source, which is basically a field state grade. Therefore, nuclear power has the lowest state grade. However, nuclear power has become relatively less costly for producing electricity by creating extremely intense radioactive fields within a reactor core. The high intensity of radiation creates economies of scale for the fuel burning cycle. It is not possible however to determine a weight or volume grade for nuclear fuels since they are fields rather like solar energy is a field and has no weight or volume. Usually though a nuclear reactor has to be fairly sizeable piece of capital equipment with a steam thermal cycle. In other words, it is not possible to use nuclear reactors in an internal reaction process similar to the internal combustion engine. Rather, nuclear cycles can only be used externally to the final use, similar to an external combustion engine. This means any engine using a nuclear fuel cycle must be bulky and heavy rather like a steam locomotive. This cuts down on nuclear power's capability for use in smaller uses such as for cars and trucks. The only other alternative is for using nuclear power to create electricity and use that for powering things. The problem with electricity is storing it, since electricity is a field type state energy grade resource. The biggest problem though is pollution. Nuclear reactor cores as well as the reactors radioactive waste can leak out into the environment creating huge environmental problems.

FUTURE ENERGY RESOURCES

In order for society to be able to transition to a new energy resource, the economy will have to adjust to alternative energy resources. Unfortunately all alternative energy resources will have much lower energy grades than currently used oil or coal. As the world's economies are forced to use less oil and more alternative energy resources, there will be a loss in economies of scale, which will adversely affect economic vitality. The following is an analysis of alternative energy resources and their energy grades.

BIOMASS

Biomass includes such energy sources as grain, forests, manure and even seaweeds or bushes, anything that is at one point biologically alive and can be cultivated. These biomass sources can be either dried and burned rather like coal or converted into ethanol or methane fuels for use in automobiles. The biggest problem with biomass sources though is that they have extremely low area grades. Just like when England had an industrial revolution going from wood to coal, where the coal was more concentrated than the wood, going in the opposite direction from coal or oil to wood would have the opposite affect. We would lose economies of scale in producing energy. There would be an industrial decline by going from coal or oil to biomass due to the loss of economies of scale. The other problem with biomass sources of energy is that they are solid state energy grades, which means they are less useful than a liquid or a gas. The best way to make biomass more useful is to convert it into a liquid or gas fuel. Such conversion processes are very costly and use a lot of energy. There is always a loss of economic efficiency when a low state grade energy resource is converted to a higher state grade energy resource, which is why high-grade energy resources are so valuable in the first place.

ETHANOL

Ethanol is the alcohol in alcoholic beverages. This is not really an energy source but an energy conversion from grain or biomass. Again ethanol suffers as an alternative energy source for the same reason that biomass itself suffers as an alternative, since producing ethanol depends on having an energy resource with a low area grade, usually biomass. No matter how efficient the technology is for turning biomass into ethanol the biomass itself is costly to harvest and use. Harvesting enough biomass for current consumption needs would require more than all US farmlands. The opportunity cost in terms of lost food production and the fuel to harvest this energy would be enormous. Much of the ethanol fuel produced by biomass would actually be used just to harvest more biomass. Right now biomass is for ethanol is harvested using oil fuels, but if oil fuels are replaced than the harvesting will use the ethanol. Once you realize this circular production need, and the fact that biomass has a low area grade, you get an idea of the enormity of the problems involved in substituting ethanol for oil.

GAS TO LIQUIDS (GTL)

Gas to Liquids (GTL) is a process for producing a liquid fuel such as methanol from natural gas. The conversion can be done at the natural gas well so that there would be no loss of economies of scale from transporting the relatively low volume grade gas. Gases high area grade could still be exploited. The gas can then be converted easily into a high weight, volume and state grade and be transported cheaply to market.

There are three problems with GTL. One is that it requires 2 or 3 times as much natural gas energy as the liquid fuel energy that is produced. That creates a loss in the amount of energy available and again the world will reach a point of resource scarcity. Another is that natural gas itself has limited supplies similar to oil. If oil will soon have low supplies then so too will natural gas. Another aspect about natural gas is the problem of entropy subsidy losses. See Reynolds (1998). In order to convert one fuel into another especially when changing one state grade into a higher state grade, there needs to be a large capital investment. However, capital machinery requires fuel to be built and run and the capital needs labor. These inputs themselves require fuel inputs. As the price of oil goes up, the cost of capital and labor inputs increases. That increases the cost of the GTL output. This means that often times energy conversions are more costly then at first thought as the economy switches from high-grade energy resources to lower-grade resources.

COAL CONVERSION, OIL SHALE CONVERSION AND HEAVY OIL

Coal, oil shale and heavy oil are all basically solid state grade energy resources. If technology can convert coal or other solid fuels into a liquid fuel, than the economy could have the advantage of a high area grade energy with a high state grade of a liquid fuel such as methanol. Again the conversion process of a solid energy to a liquid energy is subject to the entropy subsidy loss.

Oil shale is particularly difficult to convert to a liquid since it resides mostly in one small area in North West Colorado in the Piceance creak area. In order to convert massive amounts of that energy, there would need to be huge mining operations with massive capital and labor inputs in a relatively small geographical area. Putting large capital and labor inputs into a small mountainous area creates a lot of input scarcities due to a lack of services for these inputs. The inputs will be much more costly due to the high concentration needs. Compare such a small area to the East Texas oil field or Prudhoe Bay on Alaska's north slope, but multiply the capital and labor inputs by 100 fold and cut the output by at least half and you begin to get a sense of how high cost these options really are.

Heavy oil such as in Venezuela also resides in a small geographic area. Venezuela's heavy oil is in the Ornico Belt in an are only 50 miles wide by 80 miles long. Again in order to extract large amounts of liquid fuel from this area there would need to be tremendous capital and labor inputs into a small area. It is difficult to service such large labor and capital requirements in a small area adding to costs. It is as if a city the size of Manhattan were needed in the middle of nowhere. The infrastructure scarcity is enough to push the costs very high.

NUCLEAR FUSION

Nuclear fusion is a field grade energy resource. This is a low energy state grade so that nuclear fusion is mostly only useful for producing electricity. It cannot be used in an internal reaction process like an internal combustion engine, but can only be used in external reactive processes like a steam locomotive. Therefore, it cannot be used in an automobile very easily. The biggest hope though of many scientists and engineers is that nuclear fusion will be a cheap plentiful and safe source of electric power. The problem is that any fusion reactor will undoubtedly be a very costly piece of capital equipment. In fact even if the nuclear fusion fuel itself is free, the costly reactor and its undoubtedly high maintenance costs will cause the price of electricity to be at least as much as a coal fired power plant is today and probably much more. A nuclear-fired power plant needs all the capital that a coal fired power plant needs, plus the capital for the nuclear reactor itself. Where as a coal fired plant needs only a furnace chamber to burn the coal and a steam cycle generator.

Considering that nuclear fusion reactors require some of the most sophisticated high-tech capital ever developed, that capital will certainly break down quite a bit and cost a lot to build, it is very unlikely that nuclear fusion will be cheaper than coal or gas fired electricity is now. Also as oil prices go up, this will create higher capital costs due to the entropy subsidy loss of oil for capital production. This happened in the mid 1970's. As new nuclear power plants were being built during the oil price shocks of the 1970's, they became much more expensive than were originally thought. The higher priced oil caused construction costs to climb fast causing the cost of the complex nuclear power plants to climb even faster. The only advantage of nuclear fusion is that it has less radioactive wastes then conventional nuclear power and it pollutes less carbon dioxide, the green house gas that causes global warming, than coal. It is not however a perfect substitute for oil, coal or even conventional nuclear power..

SOLAR POWER

Solar power is a field state grade, which is the lowest state grade. A field state grade is low because it cannot be stored easily. Therefore solar power is cheapest when it is used as it is produced. Even that has been costly. So far the most widely heralded use of solar power is as a way to produce electricity or as a passive heating system for some houses. However, converting solar power into electricity is difficult because the solar energy is diffuse. It is difficult to obtain economics of scale for producing energy. For example in one hour, the sun can create 65 million BTU's per acre. The panels that catch this solar power must therefore be spread out just to get the energy. Coal in a mine has about 10 to 1,000 Billion BTU's/ acre. What this means is that there are large economies of scale for mining coal such that one coalmine operation can obtain much more energy per capital input than solar panels. In fact a strip mine can produce about 4 Million tons of coal per year. That is about 32 billion BTU's per hour, which can be taken from within one acre on average. Therefore, one acre of coal can produce the same energy in one hour as 500 hours of solar power. Even with lower conversion efficiencies of coal, that is still about 300 times more energy per hour from coal. Therefore the economies of scale favor coal or oil or natural gas by more than two magnitudes to solar power. This is why solar power is not really the savior of our economy. We lose too much economies of scale. Therefore, solar energy will never be cheaper than the cost of obtaining energy from those other energy sources.

HYDROGEN AND FUEL CELLS

Fuel cells take hydrogen or some other fuel such as methanol and convert it directly into electricity. The process includes a chemical absorption of hydrogen into oxygen to create water. The process produces an electric current, which can then be easily used by an electric motor for a car or any other electric use. The problem with fuel cells is the source of the fuel. While it may be cleaner and in some instances even cheaper to run an automobile with a fuel cell, there is still a problem with getting the fuel in the first place.

The most widely talked about source of fuel for a fuel cell is hydrogen. Usually the hydrogen would come from natural gas or oil or even coal. The fossil fuel is converted into pure hydrogen, stored and delivered for use. The problem with this scheme is that the oil and natural gas losses 30 to 60% of its energy content in the process of conversion. In other words, the carbon energy content of the fuel is striped away from the hydrogen energy content, which means there is less energy. Consider an oil example. If a typical barrel of oil has 1 million BTU's of energy, then after it is converted into hydrogen, it will have only 300,000 BTU's of energy. The cost of that barrel in terms of energy would be three times as much. The same is true for natural gas. You typically lose about 60% of the energy content when you convert a hydrocarbon into a hydrogen fuel.

Another problem with hydrogen is that it has a very low volume grade. At atmospheric temperature and pressure hydrogen has a volume grade of 325 BTU's per cubic foot. Again this makes it hard to store and reduces LIMM services that use it.

TECHNOLOGICAL INNOVATION

If there is less of a high-grade energy resource such as oil, then the economy must shift from a high-grade resource to a lower-grade resource. The economy will have a lower-grade input and there will be a reverse of the increasing returns to scale, causing economic decline. Technological innovation may be able to reverse such decline with higher efficiencies for the use of oil and better uses of oil alternatives.

If technology is a powerful tool to reverse a loss of a high-energy grade inputs then we should be able to observe the effect of such technology. EIA shows that the US oil use per $1,000 of GDP in 1996 dollars has gone from 1.27 Barrels in 1979 to 0.81 barrels in 1998. However, this reduction in use of oil may or may not be due to technology. Using coal and natural gas to produce electricity instead of oil is a substitution not a technological advancement. Likewise, the greater use of natural gas for heating is only a substitution. On the other hand, while substitution in many production processes and many uses for heat are possible for oil, there has not been a lot of substitution in the most vital uses for oil especially transportation.

Casual observance shows that the major uses of oil such as transportation and LIMM use have not had a lot of major technological improvements to cause large reductions in oil use. Substitutes such as electric and gas cars have not been able to over take oil fuel driven cars. 100 years of LIMM technology is still basically oil fuels technology. Mikesell (1995) proposes that in order to prove limits to growth, one needs to prove a reduction in productivity. This analysis using energy grades and economies of scale is an attempt to do that. An alternative analysis is to ask if the industrial revolution in England in the 18th century could have happened without coal, oil or natural gas. The answer is that even with today's technology, we cannot sustain the current level of GDP without coal, oil or natural gas. There certainly would not have been the great increase in productivity with out them.

CONCLUSION

It is possible that alternative energy resource will help the world's economy once oil supplies run low. However all alternative energy resources have lower energy grades than oil. As the world switches from oil to these other energy resources, the economies of scale inherent in oil's energy grades will be lost and cause the world's economies to decline. In short, the transition to alternative energies will cause lower economies of scale. That will cause economic decline. The decline in the world's economies will be simultaneous to higher inflation, creating stagflation. This is exactly what happened in the 1970's, with very high inflation simultaneously during a recession, and it will surely happen again as soon as world oil demand pushes hard against world oil supplies in the future.

References

Bronowski, Jacob (1976). The Ascent of Man, Little Publishing Co., New York.

Encarta (1998). Encarta 98 Encyclopedia, Microsoft Corporation.

Energy Information Administration (1999). Energy Statistics Annual, Washington DC.

Goldemberg, Jose, Thomas B. Johanson, Amulza K. N. Reddy, Robert H. Williams. (1987). Energy for a Sustainable World, World Resources Institute, Washington DC.

Mikesell, Raymond F., (1995). "The limits to growth: a reappraisal," Resources Policy, Volume 21, Volume 2, pp. 127 – 131.

Reynolds, Douglas B., (1994). "Energy Grades and Economic Growth," The Journal of Energy and Development, Volume 19, pp. 245 – 264.

____________ (1998). "Entropy Subsidies," Energy Policy, Volume 26, pp. 113 - 118.

Simon, Julian L. (1996). The Ultimate Resource 2, Princeton University Press, Princeton.

Chapter 10: Entropy and Diminishing Elasticity of Substitution*

ABSTRACT

In the debate over sustainable development and limits to growth, one critical issue has to do with substitutability. Pearce et. al. (1994) define "weak" sustainability as the ability to maintain a generalized capacity to produce. Solow (1974), Hartwick (1977) and Dasgupta and Heal (1979) show theoretically how weak sustainability can occur in the face of a non-renewable, non-recyclable, essential resource if the economy can substitute capital for the essential resource. However, the analyses depend on a Cobb-Douglas production function with a constant elasticity of substitution out to infinity. This paper shows that a Cobb-Douglas function eventually violates entropy and that the elasticity of substitution between any two inputs must at some point become zero. Different types of substitution are examined to show how physics and the law of entropy cause the elasticity of substitution to decline.

INTRODUCTION

Tilton (1996) suggests that in the debate over sustainable development, two paradigms have emerged. The first paradigm is where scarcity is defined in terms of its opportunity cost. In this more optimistic paradigm, mostly economists

---

* Reprinted from *Resources Policy*, Volume 25, Douglas B. Reynolds, "Entropy and Diminishing Elasticity of Substitution," pp.51 – 58, Copyright © 1999 with permission from Elsevier Science.

argue that as the supply of any given resource declines, the economy will substitute other more plentiful resources in place of the scarce resource, and the economy will continue to grow. The second paradigm is where scarcity is defined in terms of its fixed stock. In this more pessimistic paradigm, mostly physicists argue that as the supply of certain essential resources decline to nothing, the economy will also decline. Resources can be consumed and lost forever with no substitutes.

Looking at the two paradigms in terms of economics, the first paradigm is where there is a high degree of substitution possible. In other words, the elasticity of substitution between inputs is at least unity and probably close to infinity. Stiglitz (1979) suggests that the elasticity of substitution of capital for resources is sufficiently large with new technology that scarcity is not a problem, especially in the foreseeable future. If the elasticity of substitution between resources or between capital and resources is sufficiently large, then economic growth can continue without limits. The second paradigm is where there is a low degree of substitution possible so that at least one essential resource is non-renewable and has a less than unity elasticity of substitution with capital or other inputs. Therefore, one way to define the two paradigms of sustainability is to determine the substitution possibilities between important inputs. However, even if substitution between an essential resource and capital is as low as unity, Solow (1974), Hartwick (1977), and Dasgupta and Heal (1979) show theoretically how production can be sustained forever. Pearce et. al. (1994) calls this situation "weak" sustainability.

Another aspect of sustainability is technology. Lovejoy (1996) suggests that technology can change substitution possibilities over time and create efficiencies of production so there is less scarcity. Mikesell (1995) argues that there is no evidence that growth has reduced productivity. This suggests that substitutability is still relatively easy. Nevertheless an additional definition of the paradigms has to do with how forceful technology will be for increasing substitutability or production efficiency. The optimistic paradigm then has two aspects. 1) The elasticity of substitution between an essential resource and capital or an alternative resource is at least unity or greater and 2) technology will increase the productivity of resources faster than exhaustion.

While it may be true that the economy is not near any limits to substitution or near any limits to scarcity reducing technological innovation for the foreseeable future, still such limits exist. The laws of nature and particularly entropy, demand that any substitution possibility must have a declining elasticity of substitution. Even new technology cannot over come a declining elasticity of substitution between inputs. The purpose of this article is to add to the discussion of sustainable development by showing that the elasticity of substitution must inevitably go to zero due to entropy. In addition, technology cannot stop a declining rate of substitution. Consider the following questions in order to determine the characteristics of resource substitution. 1) What is entropy? 2) Are there essential, non-renewable resources? 3) Is there constant elasticity of substitution? 4) Are there such things as perfect substitutes? 5) Can resource augmenting technical change increase the productivity of a resource indefinitely?

WHAT IS ENTROPY?

According to Georgescu-Roegen (1971), entropy is where:

> Nothing is destroyed, but everything is scattered to the four winds. … the degradation of the universe covers not only energy but also material structures. As physicists put it in non-technical terms, 'In nature there is a constant tendency for order to turn into disorder.' Disorder, then, continuously increases: the universe thus tends toward chaos…

An example of the above would be a building. The glass, steel and concrete of the building continue to exist even after the building falls apart, but these materials turn into dust size particles and move to a useless state. To put it simply, entropy is degradation. A building or any kind of capital with its structural components will eventually degrade to the point that it can no longer function. The only way to slow down the degradation of a building is to maintain it. However, once the building is degraded far enough along, it must be replaced. Energy also follows this path. When energy is burned, the energy itself is not lost, however it is turned from useful work to useless dissipated heat. This is why so-called perpetual motion machines are not possible. Using any resource consumes that resource. Only if the resource is renewable, are constantly new supplies of the resource available.

Two general types of resources used as inputs into production can be defined as energy and structural resources. Energy resources create action. Structural resources are for structural components of buildings and machines. Both resource types degrade. However the rate of degradation varies when they are used such that structural resources tend to degrade much more slowly than energy resources. Since these two general types of resource inputs are pervasive and fundamental inputs into production, then it must be determined how they affect substitution possibilities. In general, since the inputs must degrade then the output too must degrade, unless more inputs are added. Therefore, entropy degrades inputs and outputs. This will affect substitution.

ARE THERE ESSENTIAL, NON-RENEWABLE RESOURCES?

The next question is whether there even exist essential non-renewable resources. The answer is yes. For any economic activity to occur at all, the economy at some point must consume energy and structural resources. Energy and structural resources can either be renewable or non-renewable. An example of a non-renewable structural resource might be concrete, which is often difficult to recycle. An example of a renewable structural resource would be wood. A renewable energy resource is like solar energy, and a non-renewable energy resource is like oil. The economy though at its current level of activity, i.e. the current GDP, is not sustainable with renewable resources. Therefore the economy depends on essential non-renewable resources. To see this, consider both energy and structural resource inputs.

Five main structural resources are metals, plastics, glass, wood and concrete. While technically metals, glass, plastics and even concrete are recyclable and therefore to a degree renewable, they require energy to be recycled. Often times a recycling process creates some loss of mass as well. However, the fact that metal, glass, plastic and concrete require energy in the recycling process makes them de facto non-renewable unless the energy they use is renewable. However, structural resources from renewable sources such as wood must be physically moved from its source location to the location of use. That also requires energy. In general structural resources require large amounts of energy to be produced. So in order to have structural resources, one usually needs energy resources. The only way for any structural resource to be renewable and indeed sustainably used is if the energy to produce or use the structural resource is renewable.

Considering energy, renewable energy resources would be such things as wood, biomass or solar energy. Estimates of biomass energy suggest that the US could only get as much as 25% of its energy needs from such sources (Smil 1983). Even if the US could get all its energy requirements from biomass energy from other parts of the world, the energy to actually transport the biomass would be almost as much as the energy of the biomass itself. Then there is the problem of what energy source the rest of the world would use. Solar energy cannot be used at night. Storing a field type energy such as solar power is difficult and requires either batteries or reservoirs that are filled up by day in order to let out the water at night for hydropower. There would not be enough fresh water in the US to fill reservoirs to store such energy. The landmass required to store seawater for hydropower would be substantial and would take up valuable farmland. The technology for storing energy in batteries has not changed in almost 100 years and is not sufficient to store the energy our economy needs to use at night. All of this suggests that the technology to sustain our current energy and structural needs does not exist.

It could be argued that once technology is better, then society can sustain current economic activity with a renewable energy resource such as solar energy. The crucial question for using solar energy to supply current economic activity is whether it can run large independent mobile machinery (LIMM). If solar cannot run such machinery, which our economy crucially depends on, then it is impossible to sustain current economic activity. However, solar energy itself depends on resources such as steel and aluminum, which in turn degrade. Batteries and reservoirs for solar energy storage would require large amounts of structural and energy resources. In addition many current economic processes such as air travel cannot be done with electricity. Considering all the inputs to make solar energy work as a sole source of energy, there would have to be some reduction in economic activity. Therefore, non-renewable resources must be regarded as essential to current economic activity.

IS THERE CONSTANT ELASTICITY OF SUBSTITUTION?

The concept used to measure the ease of substitution between two inputs of production is the elasticity of substitution, which is defined as the percentage change in the input ratio caused by a percent change in relative input prices. Assume there are only two inputs of production for defining elasticity. If there are only two inputs into production, say A and B, and if production is held constant, and if a one percent increase in the price of A relative to B causes an increase of one percent in the use of input B relative to input A, then that is a unitary elasticity of substitution. That is a Cobb-Douglas production function. If a one percent increase in the price of A relative to B causes an increase of greater than one percent in the use of input B relative to input A then the elasticity of substitution is greater than unity. If a one percent increase in the price of A relative to B causes an increase of less than one percent in the use of input B relative to input A then the elasticity of substitution is smaller than unity.

Sometimes there are more than two inputs. Kang and Brown (1981) develop a full elasticity of substitution definition that takes into account a multi input function and shows the relative ease with which any two of the inputs can be substituted. When there are several inputs into production and the output does not change, the firm can also substitute between any two inputs in order to achieve the lowest cost production of output. Consider, the relative amounts of only two of the inputs, say A and B. If the price of A and B change relative to each other, the inputs of A and B will change but so might the relative amounts of other inputs. What is of concern in a several input model case, just like a two input model case, is the relative change of quantities of input A and B given a relative change of prices of input A and B. However, in the multi-input case, when A and B change, other input quantities are allowed to change too.

An easier way to determine what is happening to the elasticity of substitution is to look at the technical rate of substitution of one input for another. For example in a two input case, if the rate of technical substitution of input B for input A is zero, then any increase in B at a constant input of A will not increase output. In that case by definition when the relative price of A increases, the relative quantity of B cannot increase for a constant output. Therefore, there is a zero elasticity of substitution when there is a zero rate of technical substitution. In this paper is shown why the rate of technical substitution always goes to zero for any two inputs of production. If the rate of technical substitution goes to zero, then the elasticity of substitution also goes to zero.

Solow (1974), Hartwick (1977), and Dasgupta and Heal (1979) consider an economy that uses only two inputs for production. They analyze the two input case to determine whether if both inputs are essential can production be sustained. Their analysis depends critically on what the elasticity of substitution is. If the elasticity of substitution is unity, then it is theoretically possible in a two input world to sustain production forever, even when one of the inputs is a non-renewable essential resource. It can never be the case that one input can substitute for another input at an elasticity of unity forever. The rate of technical substitution for any two inputs must go to zero. Therefore, the elasticity of substitution must always go to zero. Therefore there cannot be a constant elasticity of substitution greater than zero.

CAPITAL AND LABOR

To understand how entropy affects substitution consider first the substitution of capital for labor. The Cobb-Douglas equation for production with only capital and labor inputs is:

$$Q = A \cdot (L^a \cdot K^b),$$

and $$a + b = 1$$

Where A, a, b = parameters; Q = output; L = Labor input; K = capital input.

Q is a quantity of output that is produced during a specified period. That means the labor and capital inputs have to be constant for the entire period in order to produce Q. Capital and labor will not be the same input at the end of the period as they were at the beginning of the period unless they are maintained. Capital needs maintenance in order to keep up its output. Just like a car needs a yearly tune up to perform optimally, so does any capital equipment. If capital equipment is not maintained, it will produce less. Also labor needs to be maintained with rest and food. That maintenance of and by itself requires more capital.

Another way to look at the Cobb-Douglas equation is that Q is an instantaneous rate of output and L and K are instantaneous rates of input. What is missing is the instantaneous rate of maintenance of the labor and capital inputs. Usually maintenance of the inputs is not included in Cobb-Douglas or other forms of production functions since they are such a small proportion of the inputs. Maintenance inputs are usually fairly small and are often included as a part of the labor input. As long as production is not near the frontiers of the production function, the assumption of zero maintenance is close to the real world.

However if production moves along the frontier out towards the corners, then you need to separate the maintenance input from the pure capital and labor inputs. The results can be graphed. In Figure 10.1 is a capital-labor quadrant. The line along the capital axis represents the maintenance or entropy capital-labor ratio. For convenience call this the entropy ratio. The entropy ratio line represents the ratio of labor required to maintain capital at any given level of capital. If production requires 100 units of capital and the capital needs two units of labor to maintain it, then 200 units of capital will require four units of labor to maintain it and so on. Assume that more capital will be substituted for labor for a particular production output regardless of cost. As more capital is used to replace labor, each laborer will be using or in control of more capital, such that the capital-labor ratio will be higher. Entropy suggests however that capital must degrade and must either be replaced or maintained. Therefore additional laborers will be added to maintain or replace capital, i.e. to compensate for entropic loss or depreciation. Those laborers create the entropy line. An assumption of non-degradable capital is clearly a violation of the entropy law. Anyone who has had to deal with a copy machine breaking down all the time understands this.

Figure 10.1. Labor-Capital Isoquant

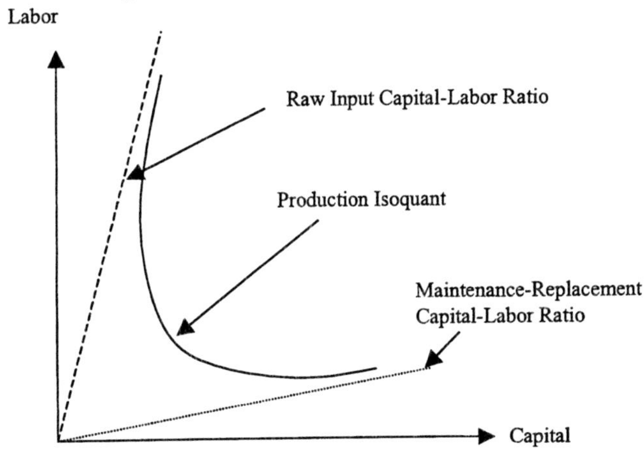

Since the entropy line has a positive slope, then the production capital-labor isoquant must asymptote to this line eventually and become positively sloped. The isoquant creates a characteristic "U" shaped curve. What happens is that the rate of technical substitution cannot be negatively sloped forever as implied by a constant elasticity of substitution. The rate of technical substitution, which is the slope of the isoquant, must eventually go to zero slope. This implies a decreasing elasticity of substitution that becomes zero at the minimum point of the U. Another example of a production function is when the elasticity of substitution for production is always less than unity. In that case, the production isoquant will asymptote to some positive quantity of labor no matter how much capital is used but will still be negatively sloped. However, the entropy ratio will start at the minimum necessary labor force above the origin and slope upwards. This again causes the production isoquant to form a U shape and reach a minimum where the elasticity becomes zero.

Often times when capital substitutes for labor, it is not just bigger or more machines that substitute for labor, but rather more complex machines. However, complex machines more often than not, have higher maintenance needs or greater degradation than less complex machines. So whether it is more capital tonnage or more capital complexity that is used to replace labor, the law of entropy holds. There is therefore a U shaped production isoquant even with more complex capital causing the elasticity of substitution to decline to zero. Increasing complexity is just as apt to need replacement or maintenance as increasing size or quantity of capital.

In some instances capital can be so run down that it needs to be replaced or at least some parts of the capital stock need to be replaced. One necessary input for replacing capital is structural resources, which degrade due to entropy. Therefore, as structural resources within capital degrade, then so does the capital itself. This is depreciation. Depreciation and maintenance both together can be considered as the entropic decay of capital. Therefore, whether capital is maintained or replaced or parts of it are replaced, the end result is a declining elasticity of substitution. It is not possible to maintain or replace capital without using more labor. The entropy line does not necessarily have to be a straight line, but it is positively sloped at all times.

On the other hand, if more labor is used to substitute for capital, again limits are reached. In order to increase labor, labor needs to have food. Food is produced using labor and capital. Even the most primitive of tribes uses at least some capital to produce food. Therefore as an economy attempts to use more labor to replace capital, it needs at least some capital per laborer just to feed each laborer. There is therefore a maximum labor-capital ratio starting at the origin and proceeding out ever so slightly with a positive slope close to the labor axis. This is called the raw input capital-labor ratio. It is the necessary minimum where labor with a minimum amount of capital can actually contribute to production. The raw input ratio does not need to be a straight line but is positively sloped at all times. The production isoquant shown in Figure 10.1, must then asymptote to the positively sloped raw input line. The production isoquant will asymptote to both the raw input ratio and the maintenance replacement ratio creating a "C" shape. The technical rate of substitution of labor for capital will reach a vertical slope and therefore a zero elasticity of substitution at the furthest left-hand point of the "C".

ENERGY AND CAPITAL

Apostolakis (1990) shows that in the short run energy and capital are complements, that is there is a zero elasticity of substitution between them, but in the long-run they are substitutes. If energy and capital are substitutes even if it is only in the long-run, then it may be possible to use less energy by building more capital. Advocates of sustainable growth through efficiency look to energy efficiency by way of better technology or more capital as a way to sustain growth. For example a firm could use trucks with fuel injected V-6 engines rather than a standard V-8 in order to get similar performance and power, but with less energy. However these more complex engines endure higher entropy degradation and so usually require more maintenance or quicker replacement or both. So again there is a greater maintenance-replacement need as capital use increases due to the entropy law. Every time capital is replaced or maintained, it requires a certain amount of energy input.

Looking at capital and energy in particular, the more capital you use to replace energy, the more maintenance or replacement of capital is needed. However, it takes energy and capital to maintain capital and to produce replacement capital. This suggests that there is a maintenance-replacement ratio of energy for capital, which is the entropy line for capital similar to Figure 10.1. The maintenance replacement ratio of energy for capital can be modeled in the same manner as the ratio for labor and capital. As the energy-capital isoquant moves to greater use of capital to substitute away from energy, then eventually the isoquant will asymptote to the entropy line and create a "U" shape. That means the elasticity of substitution goes to zero.

When it comes to substituting energy for capital in the long-run, you can use more energy with very simple capital in order to substitute more energy for less capital. However, in order to substitute energy for capital, there will be another boundary that is run into. It is impossible to produce, transform or transport energy without capital. Therefore, there must be a minimum amount of capital to produce energy and have it available for any production process. This is called the transformation capital-energy ratio. As energy use increases, there is a necessary minimum increase in capital to be able to provide the energy. There is therefore a transformation ratio line for energy that starts at the origin and goes with an ever so slightly positive slope out from the energy axis. Again, the energy capital isoquant must asymptote to this line creating a "c" shaped curve, similar to Figure 10.1. This implies that the elasticity of substitution of energy for capital goes to zero.

ENERGY AND LABOR

On the energy-labor isoquant, Labor can be substituted for energy and vice-versa. However, the only way to use more energy is with the use of more capital since it requires capital to produce energy as explained in the energy capital substitution scenario. As energy is substituted for labor, capital must be simultaneously substituted for labor. However, since capital is subject to entropic decay and must be maintained by labor, then there is a minimum labor-capital ratio that must be maintained the use of capital is increased. This minimum labor-capital ratio can be transformed into a minimum labor-energy ratio, since energy expansion needs capital expansion and capital expansion requires a minimum expansion of labor for maintaining entropic decay. Thus on the energy-labor quadrant, there is a positively sloping input-entropy line that proceeds slightly out from the energy axis, which is the maximum energy-labor ratio needed to have production occur. Any energy-labor isoquant must asymptote to this line and so must have a declining elasticity of substitution that reaches zero. This gives a figure similar to Figure 10.1.

On the other hand, if labor is substituted for energy, then less energy intensive capital is used. However raw labor needs food and food comes from capital as well as energy. Also since capital entropically degrades then there must be at least some maintenance-replacement energy input for every piece of capital that labor uses. Therefore there is a positively sloped maximum labor-energy ratio line, again called the input-entropy line, that starts at the origin and proceeds slightly from the labor axis. The labor-energy isoquant must asymptote to this positive slope such that eventually the elasticity of substitution becomes zero.

LAND SUBSTITUTES

A classic example of diminishing returns to production is where a farmer with one acre of land hires additional laborers. At first, extra laborers create synergy and there are increasing returns per marginal increase in labor, but then marginal increases in labor add less production due to diminishing returns. Eventually production actually declines with marginal increases in labor, because the laborers get in each other's way since there are too many of them to fit on the finite piece of land. This implies a crest on the production hill that starts close to the labor axis and goes parallel to the land axis. A single production isoquant for land and labor in that case must have a "U" shaped curve similar to Figure 10.1 in order for there to be such a crest in the production hill. That further implies that the elasticity of substitution of labor for land goes to zero.

This same story goes for capital and energy substitutes for land. If more capital is used, then there is an increase in the capital per acre, until capital starts interfering with itself and again a situation like Figure 10.1 arises. Since energy is used in conjunction with capital, a land/energy isoquant will also be "U" shaped.

On the other hand, if more land is used to substitute for less labor or less capital, then there is a different problem. There is a time constraint on the use of labor or capital such that these inputs can only work a finite amount of land in one growing season. These inputs cannot work more than a certain maximum amount of land since there is not enough time to work more land then that maximum. This maximum amount of land that capital or labor can work can be graphed as a positively sloped maximum usefulness line that proceeds from the origin of the land/labor quadrant out close to the land axis. The land/labor or land/capital isoquant will not asymptote to the maximum usefulness line, but will cross it at exactly the point where the isoquant tangent is parallel to the land axis. Therefore the isoquant has an elasticity of substitution of exactly zero when it intersects the maximum usefulness line.

POLLUTION

Pollution can be considered to be an input to production. By polluting more, firms produce more. Pollution is also an input to enjoying our lifestyle like when we drive our cars. As an input, then, it should be possible with constant technology to be able to cut down on pollution by substituting other inputs for the pollution input. For example smoke scrubbers on smokestack plants are a way to substitute capital for pollution. There are also ways to substitute labor for pollution like for example using mass transit, which usually takes more commuting time, i.e. labor, but saves on car pollution.

However, even with pollution, there is only so much capital and labor that can be substituted for pollution. For example, when capital is used to substitute for labor, the capital itself uses pollution inputs for production. Since the capital degrades it must be maintained or have parts be replaced or have the entire mechanism be replaced. This creates a maximum capital-pollution ratio similar to the capital-labor ratio. Or if pollution is replaced with labor, like using mass transit, there is still some pollution from mass transit no matter how much people use mass transit.

ARE THERE SUCH THINGS AS PERFECT SUBSTITUTES?

Inputs into production other than labor can be narrowed into two main resources, energy and structural resources. It is easy to see that in a lot of cases one energy resource, e.g. nuclear power, can be substituted directly for another, e.g. coal. Also it is possible for one structural resource, e.g. steel, to be substituted for another, e.g. wood. Gordon et. al. (1987) for example look at metal substitution. However, the concept of perfect substitutes is more complicated in the real world. Reynolds (1994) has defined energy grades that show that even when one energy resource has the same energy content as another, other characteristics such as the physical state of the resource can make one resource easier or harder to use.

Consider an example of substituting nuclear power for coal. In order to use nuclear power in place of coal, a firm needs to build a nuclear power plant, which uses highly refined steel and alloys of much higher quality than a coal fired power plant. It is however necessary to produce steel using coal. Therefore whenever nuclear power is substituted for coal, firms still need to use at least some coal just to be able to produce the steel in order to build the power plant and maintain it. This is just one simple example of a complex relationship in the real world between many and varied energy and structural inputs. While it is conceivable to substitute coal for nuclear power, such that coal is a perfect substitute for nuclear power, it is almost inconceivable to substitute completely nuclear power for coal due to the need to use coal to produce steel. In general, it is expected that higher-grade energy inputs will more easily and completely substitute for lower-grade energy resources, but lower-grade energy resources will not easily or as completely substitute for higher-grade energy resources.

When a firm tries to substitute structural inputs such as steel for wood, similar problems come up. It takes energy to produce steel and energy to transport wood. Interestingly there is only one renewable energy resource that can actually be used to produce steel and that is wood. Therefore in the long-run, the only way to replace wood with steel is by using wood as a carbon source for the steel. In general an economy with more complex inputs will usually have fewer substitution possibilities for structural and energy inputs.

Looking at economic history, the usual view is that technology creates growth. However, upon closer examination, it has often been higher-grade energy resources and higher-grade structural resources that have created growth, not just technology alone. For example the change from coal to oil use was a change to a higher-grade energy resource and the change from iron to steel was a change to a higher-grade structural resource. A complex economy, using many high-grade resource inputs will have fewer substitution possibilities to replace any one of its inputs, then a less complex economy that relies more on low-grade inputs. It is easier to substitute higher-grade resources for lower-grade resources, than to substitute lower-grade resources for higher-grade resources. Therefore, economic history is just as much a history of increasing resource grades as it is a history of increasing technology. However, if an essential high-grade resource is scarce, then it becomes increasingly difficult to substitute for it. In general it may be the case that a high-grade resource is a perfect substitute for a low-grade resource, but it may not be the case that a low-grade resource is a perfect substitute for a high-grade resource.

CAN RESOURCE AUGMENTING TECHNICAL CHANGE INCREASE THE PRODUCTIVITY OF A RESOURCE INDEFINITELY?

It is often suggested that even if society cannot substitute indefinitely one resource for another, that technology will improve to the point that new substitutes will emerge for relatively scarce resources. One way to look at this is to ask can technology create infinite productivity from a resource. Consider energy. Can an electric power plant ever convert more than 30% or 40% of the input fuel into electricity? In fact physical entropy laws of heat and mass transfer restrict the efficiency of any thermal process. For example a Carnot cycle at atmospheric pressure between ice water and steam is 27% efficient. With an imperfect process you get even less engineering efficiency, but in no way can you get more. Other cycles are also limited. See Van Wylen and Sonntag (1978).

Consider structural resources. The molecular make up of any material will determine its relative strength, diamonds being very hard and lead being fairly soft. The hardness and strength of diamonds is due to the compactness of the carbon molecules. It is possible to determine any materials structural integrity, strength, and lightness based on how compact its molecules are and their molecular weight. No technology can change that strength. However, there are always new materials and alloys. On a molecular level though, alloys will only be as strong as their potential to have their molecules form compactly together and there are limits to molecular compactness. New materials such as plastics are also limited in how compact the molecules can be. No matter what new materials are discovered, technology will reach a limit of structural integrity and strength.

In addition structural resources depend on energy to be produced. Again it is possible to determine a minimum energy requirement to physically melt a structural resource or in any other way convert it. Even wood requires a minimum energy to transport the material. In general, laws of nature will limit the productivity of energy and structural resources. Entropy limits the productivity of resource inputs and technology.

CONCLUSION

Some may say that we are not near any limits of substitution such that diminishing elasticities of substitution may as well be infinitely away from the portion of the production isoquants we currently live in. However, it is always possible that we are closer to such limits than we realize. Technology can also change isoquants, shifting them further and further in, so that we get further away from zero elasticities of substitution. However, entropy itself limits how far these isoquants can shift inward, such that no matter how good technology is there is an entropy limit on the minimum of an isoquant. Nevertheless, no matter how far away technology takes us from zero elasticities, they are still out there. The question remains, are we still in an area on our current production isoquants where technology can dominate entropy or are we getting close to entropy limits. It is important to at least recognize that elasticities of substitution must and do decline in the same manner that economists already recognize in the law of diminishing returns.

The main implication of diminishing elasticities of substitution is that there are limits of substitution. Society cannot count on being able to cut down on pollution or energy use indefinitely with mere substitution alone. Eventually the only way to reduce pollution or energy use below a certain point is by actually cutting industrial production or consumer consumption itself and thus reducing GDP and our standard of living.

REFERENCES

Apostolakis, Bobby E. (1990). "Energy-Capital Substitutability/Complementarity: The Dichotomy," Energy Economics, Volume 10, pp 48 – 58.

Dasgupta, P.S. and G.M. Heal (1979). Economic Theory and Exhaustible Resources, Cambridge University Press, pp. 196-207.

Georgesu-Roegan, N. (1971). The Entropy Law and the Economic Process, Harvard University Press, Cambridge, Mass..

Gordon, R.B., Tjalling C. Koopmans, William D. Nordhaus and Brian J. Skinner (1987). Toward a New Iron Age? Quantitative Modeling of Resource Exhaustion, Harvard University Press, Cambridge, Mass..

Hartwick, J.M. (1977). "Intergenerational Equity and the Investing of Rents from Exhaustible Resources," American Economic Review, Volume 67, no. 5, pp. 972 –974.

Kang, H. and G.M. Brown (1981). "Partial and Full Elasticities of Substitution and the Energy-Capital Complementarity Controversy," in Modeling and Measuring Natural Resource Substitution, eds. E.R. Brendt and B.C. Field, MIT Press, Cambridge, MA, pp. 81 –89.

Lovejoy, Derek (1996). "Limits to Growth?", Science and Society, Volume 60, no. 3, Fall, pp. 266 –278.

Mikesell, Raymond F. (1995). "The Limits to Growth a Reappraisal" Resources Policy, Volume 21, No. 2, pp. 127-131.

Pearce, D. W., G.D. Atkinson and W.R. Dubourg (1994). "The Economics of Sustainable Development", Annual Review of Energy and the Environment, ed. Socolow, R.H. Volume 19, pp. 457 – 474.

Reynolds, Douglas B. (1994). "Energy Grades and Economic Growth," Journal of Energy and Development, Volume 19, Number 2, Spring, pp 245 – 264.

Smil, Vaclav (1983). Biomass Energies: Resources, Links, Constraints, Plenum Press, New York.

Solow, R.M. (1974). "Intergenerational Equity and Exhaustible Resources" Review of Economic Studies, Symposium on the Economics of Exhaustible Resources, pp. 29-45.

Stiglitz, J.E. (1979). "A Neoclassical Analysis of the Economics of Natural Resources" in Scarcity and Growth Reconsidered, V. Kerry Smith Ed., Resources for the Future, Baltimore, pp. 36 – 66.

Tilton, John E. (1996). "Exhaustible Resources and Sustainable Development, Two Different Paradigms" Resources Policy, Volume 22, Nos. 1/2, pp. 91-97.

VanWylen, G.J. and R.E. Sonntag (1978). Fundamentals of Classical Thermodynamics, SI Version, 2nd Edition, John Wiley and Sons, New York pp187.

Chapter 11: Soviet Economic Decline: Did an Oil Crisis Cause the Transition in The Soviet Union*

INTRODUCTION

If Marxist-Leninist Communism was so inefficient, then why did it last in one form or another for over 70 years in the Union of Soviet Socialists Republics (U.S.S.R.)? Did it last so long because inherently it provides a good economic system or was it not as inefficient as we are made to believe, but that some other circumstances caused its demise? I believe the latter. The Soviet system was not spectacularly inefficient and was an adequate, though far below average, system for running an economy. As long as there were plenty of resources, the Soviet economy could continue running. However, rather like a gas-guzzling car, the planned economies became spectacularly inefficient in the face of resource scarcity, particularly oil scarcity. A planned economy is not flexible enough to reallocate quickly scarce resources that suddenly decline in production.

I believe it is no coincidence that Soviet oil production declined by 30% from 1988 to 1992, BP (1996), precisely during the time when the break up and capitalist transformation was occurring. Furthermore since 1992, Russian oil

---

* "Soviet Economic Decline: Did an Oil Crisis Cause The Transition in The Soviet Union?" by Douglas B. Reynolds, *Journal of Energy and Development*, Autumn 1998, Volume 24, Number 1, pp. 65-82, Copyright © ICEED 2000; with permission from International Research Center for Energy and Economic Development (ICEED), 850 Willowbrook Rd., Boulder, Colorado 80302, USA

production has declined an additional 20% or more and during that time Russia's recession has worsened. Most analysts see cause and effect during the Soviet transition as being one of the transition of the economy causing the oil production decline. I however believe it is the opposite. It was the oil production decline that caused the final break down of the Soviet economy. The collapsing Soviet economy then caused the break up and political change of the Soviet System. Certainly, there were many factors that played a role in the fall of the Soviet Union. However, there is no bigger fly capable of breaking a camel's back than oil.

In 1973 and 1980 when the West went through huge economic problems due to oil price shocks, the Soviet Union did not appear to have nearly as large of problems. The West went through an incredible oil crisis, which was discussed and analyzed to incredible lengths, yet the Soviet Union continued relatively unfazed. This was because the Soviets had plenty of oil at that time to fulfill their own needs and even enough surplus to export. Even during the so called Soviet "oil crises" of 1977-1982 and 1982-1988, Gustafson (1989), the Soviet Union did not have decreasing oil production except for a slight decrease in 1984. During both periods the Soviet Union could cover consumption and exports and the economy was not greatly affected. However, by 1989 that had changed. Finally the Soviet Union faced the same economic crisis that the West had already endured in order to allocate its oil more efficiently.

In 1989, official Soviet oil consumption was higher than the former Soviet Union's entire oil production of 1993 (BP 1996). Also in 1989, the Soviet Union officially exported 30% of its oil, and much of that export was to the Eastern European satellites. Together the Soviet Union and Eastern Europe consumed roughly 20% more oil in 1988 than they produced in 1993. The only way that the Soviet Union could have possibly survived such an oil production decline crisis, not to mention a complete loss of oil export revenue for hard currency, as well as cut its internal consumption, and cut off Eastern European exports, would be to shift to a market economy. The Soviet Union did exactly that. First, the Soviet's cut off Eastern Europe from receiving cheap Soviet oil and forced them to pay in hard currency prices. Then when that was not enough the Soviets themselves had to allow internal oil prices to rise. This forced the Soviets to follow the same transition that Eastern Europe did. Indeed, Soviet and post Soviet oil consumption declined a staggering 50% from 1985 to 1995. However, the transition was not by choice but by brute economic force.

In this paper will be shown first the theories for the fall of the Soviet Union, then why oil is important to an economy, and then how oil affects economic growth. Specific facts about the Soviet oil crisis and Soviet and Eastern Europe economic decline will be shown. Finally, the paper will explain why Soviet oil production declined, which was not because of inefficient Soviet planning, but rather because of scarcity.

THE COMPETING THEORIES

The classic economic explanation for the decline and break up of the Soviet Union is given by Kotz and Weir (1997). They argue that it was economic stagnation that precipitated political changes towards democracy. I agree with their analysis but add one more element to the economic sequence of events, which is an oil crisis. Other variations of the economics of the Soviet transition can be found from Krueger (1993), Kafouros (1996), Ellman and Kontorovich (1992), and Osband (1992). The sequence of events is as follows:

1. Even though a planned economic system is relatively inefficient, it is still able to obtain high rates of growth by merely moving from an agrarian economy to an industrial economy. During industrialization, it is a relatively simple feat to borrow the best technology from other industrialized countries and start using it. However, as a greater percent of a work force becomes factory workers rather than farm workers, there is less potential for economic growth based on industrialization alone. Further growth depends on innovation. Thus, the Soviets were faced with the transition from extensive growth to

intensive growth; i.e. growth based on technology not mere industrialization alone. Not only the Soviets, but many other societies also faced this problem. See Ellman (1986). Unfortunately for the Soviets their economic system was especially weak at innovation. In general, growth based on innovation rather than transformation is slower, particularly for a planned economic system. By the mid 1970's the Soviet Union started stagnating.

2. As stagnation in the Soviet economy continued, the people and even the party leaders began to become impatient. They wanted faster growth. Not only was the desire for growth caused by the party wanting more income for its members but also to quell discontent in the population. The perceived need for more economic growth allowed Gorbachev to rise to power in 1985 on a platform of reform. Gorbachev followed through with his platform by instituting "*perestroika*," "glasnost," and greater economic freedom.
3. The new economic and political openness created even greater economic inefficiency due to the double direction that these policies went. Some segments of the economy were freed; namely wages, while other segments of the economy were still being planned or regulated; namely prices, Osband (1992).
4. The greater freedom *perestroika* engendered for some aspects of society created a changing society that had gradually gotten used to more freedom and free speech and thought. Soviet society was changing quickly. The Soviet People were seeing more and more how they were behind the West, and they were more boldly expressing their discontent, Dallin (1992)
5. The partially free, partially planned hodgepodge of economic policies made the economy worse rather than better, so the economy declined even further.

6. As the economy declined still more, greater political discontent and fractionalization within and outside of the party occurred. The discontent induced an even more dichotomous policy: trying to free the economy more and at the same time controlling it more.
7. Eventually this bad economy, necessitating bad policies within the framework of a command system, caused an even worse economy. Finally the sinking economy produced enough discontent that the political system broke down and the Soviet Union broke up.

An alternative theory of the collapse of the Soviet Union has to do with military buildups, see Brown (1992). In this model, we have the same economic stagnation as before, but it is made worse due to more and more resources being poured into the defense industry. As defense gets a bigger piece of the Soviet pie, there is less investment in capital. The total pie gets smaller, leading to economic and then political collapse. America's own defense industry has made claim that President Reagan's military expansion helped caused this. The theory suggests that the US military buildup induced a Soviet escalation of its own military power. The Soviets, trying to keep up with the West, starved their economy of needed investment, speeding up their economic decline.

These general explanations however do not explain why the break up occurred exactly in the late 1980's and early 1990's. The idea that wages were allowed to rise but prices were kept constant does not take into account the thriving black market that occurred during this time. See FBIS (1990). Nor does the theory explain why it was that the Soviet authorities started the process of glasnost in the first place and choose to raise wages all of a sudden in the late 1980's. The reasoning becomes circular: Soviet mismanagement caused Soviet mismanagement. At any rate, since a black market existed throughout the Soviet economy, we cannot use Soviet planning as the singular cause for its monumental decline. Black markets are powerful economic structures that can overcome the most fool hardy of policies. Furthermore, the Soviet and Eastern European economic decline did not start in earnest until right at 1989, and all factors mentioned existed before 1989. For example the Soviet military had always taken a larger percent of GDP than the United States did for defense expenditures, even as far back as the Russian revolution. Furthermore, the Soviet military was always behind the West technologically, so they knew they could never gain an advantage nor even close the gap. Hence, we are left searching for an alternative reason for the events of 1989 to 1991. There must be a more definitive cause and effect scenario for how and why the collapse of the Soviet Empire happened at the exact time that it did.

THE ROLE OF OIL

I believe a more fundamental problem in the economies of Eastern Europe and the Soviet Union was the decline in Russian oil production. Russia was the main oil producer for the communist countries, and its production decline pushed the Soviet and Eastern economies over the brink of disaster, causing the collapse. What is more, the oil production decline did not stop with the break up but has continued to this day, exacerbating the economic problems of Russia and the former states. Certainly much of the economic stagnation in the 1970's and 1980's was caused by the end of industrialization and the lack of innovation. Possibly military defense spending did play a role in further economic collapse, and as Dallin has said, Gorbachev and his policy of glasnost were a big factor in changing Soviet society since this allowed people to feel freer to react politically and express their desire to change communist rule. However, like Weir and Kotz, I believe the final cause of the break up was economic decline. In addition, I believe the final decisive economic event-the final straw that broke Soviet economics-was the decline in oil production.

Consider first how valuable oil is as an input into production. Gisser and Goodwin (1986) show that the price of oil significantly affected the growth rate of the US economy. Boyd and Caporale (1996) show that not only oil prices affect economic growth but even the uncertainty of oil prices can affect growth. Trying to make a transition from using oil, to using alternative energy resources such as coal or gas, there is bound to be a price rise for oil and such a price rise would have to create economic problems. It did so for the West. It would also have to be true for the Soviet Empire.

In order for the Soviet Union and Eastern Europe to overcome the loss of oil, their economies would need new technology, more innovation, and greater allocation efficiency for their oil resources; and only a rise in prices could make that happen. In the final years of the Soviet Empire, exactly from 1988 to 1992, the command economies were being forced either to use less oil, to switch to alternative energy sources, or to do both, and in a very short amount of time. This was a fast sudden collapse. The price of oil had been artificially set much lower than the world price and much lower than its scarcity value within the communist system (Balabanov and Deitz 1991). This low price coupled with virtually unlimited supplies up to the 1980's, subsidized the Soviet and Eastern European economies. However, once the oil became scarce, both due to continued small economic growth and due to languishing supplies, internal oil price had to go up. This created an oil crisis.

Consider the history of the Soviet economy all the way up to and past its break up. In particular compare the Soviet and US economies in relation to oil consumption. I look at oil consumption statistics rather than price statistics because the Soviet Union kept internal oil prices artificially low and constant. In addition, oil consumption statistics within the Soviet Union are generally correct but not necessarily very accurate due to the desire to show high growth in every sector of the economy. It is therefore not possible to econometrically relate oil statistics to economic growth, especially since other changes in the economy were occurring simultaneously. Never the less according to the CIA, Soviet oil consumption was increasing at roughly a 7% per year rate and oil production at roughly an 8% per year rate from 1960 to 1975 (CIA 1985). During that time the CIA estimates that the Soviet Union had a roughly 6% rate of economic growth. From 1975 to 1980, the Soviet Union had a much lower 4% per year increase in use of oil and only a 3.5% increase in yearly production. This is the time when Soviet growth fell to 2.6%. From 1980 to 1985, the Soviet Union had no increase in oil consumption and a slight decline in oil production. At that time economic growth fell to 1.8% per year according to CIA estimates. See CIA Directorate of Intelligence (1988). Finally, from 1988 to 1992, Soviet and post Soviet oil production collapsed, as did consumption. At that time both Eastern Europe and the Soviet Union and the Former Soviet Union went into severe economic decline. What is striking about this period is that production declined first then consumption not the other way around. If one was to expect an economic crisis causing an oil production decline, then the reverse would happen.

In fact growth and decline of an economy and the growth and decline of oil consumption go hand in hand. Consider the US. From 1960 to 1973, the US had a 7% growth in oil consumption and a 5% economic growth rate. From 1973 to 1975 the US consumed roughly 6% less oil and had a recession with a 6% economic decline due to the oil price shocks. From 1975 to 1979 the US oil consumption rate increased at about 6% per year while there was simultaneously a 5.5% per year growth rate. Then starting in 1979 another set of oil price shocks started more oil conservation. From 1979 to 1982, US oil consumption declined at an incredible 5% per year, with no growth in the economy during that period, and a major recession (EIA and Economic Report to the President various years).

The US experience as well as Western Europe's points directly to a link between oil availability and growth especially in the short run. In Western economies, availability of a resource is determined by price. In planned economies, only quantity determines availability. High oil prices definitely affected the US and other free economies. The same must have been true for the Soviets. In the West, oil scarcity was visible due to the skyrocketing oil prices. This induced oil conservation, at the expense of going through two major recessions and lower overall growth rates since 1973. However, the Soviet Union had to make the same kind of adjustment. They had to have a similar short run shock in order to more efficiently use oil. They had to have an oil crisis. Their oil crisis though was much less visible due to regulated prices, the confusing price structure put on everything, and the hyperinflation that the government resorted to during the transition. Still the Soviet's had to go through the same oil conservation transition that the rest of the world had already made in the 1970's and early 1980's, only they had even less flexibility than the US for adjusting to input scarcities. Designs for fuel-efficient capital and transportation as well as the implementation of fuel-efficient methods of production and transportation could not possibly be developed and implemented as fast in a planned economy as it could in a free market economy. This is why many of the Soviet party elite in the final days pushed ever harder for more openness and especially freer markets.

This brings up the peculiar dichotomy that existed between East and West during the tumultuous 1970's and 1980's. In the mid 1970's from 1973 to 1975, almost the entire world was having severe economic troubles due to the oil price shock. The Soviet Union and Eastern Europe though seemed relatively well off all be it with some stagnation during this time. Their economies appeared less traumatized. There was no rioting or policies to reduce military outlays drastically. By contrast, the West was full of energy talk and was grasping for ways to overcome energy shortages. The US even hints at a military strike against Saudi Arabia in order to occupy the oil fields, TIME (1975). No such ideas were coming from Russia. Then in the late 1970'as the Western economies seem to be back on track, the Soviet Union staggers a bit economically.

In the early 1980's, the West again went into disruption with another round of oil price shocks. All the leading countries endured a recession. Yet the Soviet Union, while somewhat staggered, seemed to ride through the 1980's price shocks with no serious disruptions. In fact, they were helped by the oil price increases so that they could trade oil to the West for Western goods and technology. Only Poland and maybe some other Eastern European countries were affected at this time. Finally, the end of the 1980's came with most Western economies flying high, but with Eastern Europe and the Soviet Union experiencing economic and political collapse. We are left to wonder why the East was insulated from the same oil troubles that the West had experienced, only to collapse so suddenly out of thin air when the Western economies were doing so well.

The differences are much clearer when we consider oil consumption patterns. Every time the West went into a recession during this period, it was because it was forced to consume less oil due to high oil prices. However, the communist planned economies had enough oil during those times. Finally, however, the Soviet Union and Eastern Europe were forced to use less oil at the end of the 1980's. Their oil crisis finally erupted. Now we can understand the dichotomy between East and West. The Western recessions occurred during times of decreased oil use. The same was true for the East. However, the Western periods of transition were at different times then the Eastern periods because the two sets of economies were virtually cut off from one another. The East was dependent on Soviet oil while the West was dependent on Middle Eastern Oil. The oil scarcity of East and West happened at different times and this is why the growth rates were affected at different times. The most severe decrease in oil production in the East started in 1989, which undoubtedly had a huge stagflationary affect on the Eastern economies.

If we consider the time period of the mid-1980's to the mid-1990's, Soviet and former Soviet oil production declined from a high of 12.66 million barrels per day in 1987 to a low of about 7 million barrels per day in 1995 (BP 1996). Eastern Europe and the Soviet Union combined consumed about 10 million barrels of oil per day or more in 1987. Today they have managed to cut consumption to roughly 5.5 million barrels of oil per day. That is almost a 50% cut in ten years. If, however, the command economies had continued to depend exclusively on Soviet oil, then they would have either had to import oil at the rate of about 3 million barrels per day or make their economies more oil efficient by 30%. Could they have managed that? The Eastern economies had virtually nothing of value to export to pay for so much oil. On the other hand, even the Western economies during the height of the second oil crisis could only manage a 20% reduction in oil usage while they became more oil efficient. Yet that 20% reduction caused monumental problems for the West. How could planned economies have faired better? The planned economies would have had to experience an even more painful transition than the West. In fact, they did. They freed their economies, however, it was virtually impossible-not to mention hypocritical-to follow the tenets of any communist ideology with government ownership of everything and then suddenly change to a free market economy.

Therefore, in order for the Soviet Union and Eastern Europe to become more oil efficient, as quickly as possible, they were forced to change to a market economy and then democracy. Furthermore, intense discontent of the population due to the oil crisis and economic downturn, as well as a lack of military resources for internal control, made it virtually impossible to hold on to power. To summarize, the period 1988 to 1992 was the world's third oil crisis in 20 years. This one brought down the powerful Soviet Empire.

Unfortunately, for the communist party, the free market push did not go fast enough. From 1988 until 1992, Soviet oil production fell an astounding 9% per year. These were the exact years of the fall of most of the communist governments in the Soviet sphere as well as the break up of the Soviet Union itself and the fall from power of the Soviet communist party: the economy of the Soviet Union, the post-Soviet states and Eastern Europe crashed by the same magnitude of force as the oil production crash. The exact level of economic decline in the Soviet and post-Soviet world is difficult to say, given the inexact statistics and currency changes. Never the less, it would be difficult to refute the fact that these two incredible events of economic decline and oil production decline occurring at exactly the same time were not related. Furthermore, given the magnitude by which the Western economies were affected by their oil crisis, the Eastern economies would have been even more affected since they could not adapt as easily as the West had and because they had to reduce oil consumption even more.

The most startling aspect of the oil crisis in the Soviet Union after 1988 is that it is oil production that declines first and then oil consumption. If one were to have expected oil production to decline due to an economic crisis, then the economic crisis would first cut consumption significantly long before it would have affected production. This did not happen. First Soviet production began declining then oil shortages appeared and then came an economic crisis finally followed by huge consumption declines. If the Soviet Union had transformed itself in 1980 or 1975, then there would have been consumption declines but no production declines. It was opposite of this in the transition of the early 1990's.

I believe the cause and effect of the Soviet transition is politics reacting to oil production, not oil production reacting to politics. In the West, all the evidence suggests that politics reacted to oil prices. It had to be the same in the East. The Soviet economy was not flexible enough to cope with diminishing oil supply and higher oil prices. In the closed Soviet system, the price of oil was set artificially low relative to the outside world and to the true value of oil. When the supply production declined, oil prices had to go up inside the closed system. It was, however, hopeless to try to save a planned economy since it could never adjust to the strain. The economic problems of reducing the use of so much oil were enormous. If we acknowledge that there was indeed an oil crisis in the Soviet Empire at the time of its break up, then we must reexamine the sequence of events as outlined in the competing theories section above. I believe that point 3 of the classical model of the Soviet decline should be changed, or at least a new point should be added between points 3 and 4. The new point would explain how the economy had to adjust to an oil crisis brought on by the decline in Soviet oil production. This decline was the final and decisive push that broke up the Soviet Union and dismantled the Soviet Communist party.

THE CAUSE OF THE SOVIET DECLINE IN OIL PRODUCTION

Most analysts claim that the cause of the Soviet oil production decline was a combination of old technology, a lack of investment into the oil industry, and poor management. In addition, the break up of the Soviet Union itself and the problems of the economic transformation are also considered important factors. See Reinsch et. al. (1992). However, there is evidence to suggest that the Soviet oil production decline was natural and would have happened no matter what the Soviet Union did.

Consider first the idea of technology. It is suggested that because the Soviet Union had out-of-date oil technology that production decreased. However over time, even in a closed system such as the Soviet Union, information about technology and technology itself must increase. Soviet technology, although behind the West, was always getting better, not worse. So, consistently better technology of and by itself cannot cause lower production. Even the US since 1970 has had consistently better oil production technology and stable real prices; yet oil production has never the less declined.

Another possible cause of oil production decline is mismanagement. Again, however, the only way for mismanagement to cause oil production decline is if the level of management efficiency declines over time. There is no reason to believe that management was much better in the 1960's when oil production skyrocketed then in the 1980's when it stagnated. There is always incentive to produce more oil with a given set of inputs, whether under a communist or capitalist system. The incentive in a command system was to climb the party ladder or to become a manager or even to get paid more since the Soviets did have incentive pay systems. So incentive was always there to manage the resources the best way possible and to produce as much oil as possible. Unfortunately with oil, it is possible to produce too much. If managers over-produce oil when oil fields are initially put into production, which is suspected in Russia, then production can decline faster in later periods. In other words, Russia may have tried to produce oil too fast in the 1970's and 1980's. The over-production caused a sever decline in later oil production in the late 1980's and early 1990's. This point could be a factor in Soviet oil decline; but it cannot explain everything. It is a long-run phenomenon. It did not occur over-night.

Consider, if the decision to extract too much oil in the 1970's and 1980's was the cause of the 1990's decline in production, then it cannot be the transition problems of the late 1980's that caused the oil production decline. In other words, we cannot blame the oil production decline that started in 1988 on a few years in the late 1980's. In addition, extracting oil more slowly in early periods cannot prevent an eventual oil production decline; it can only delay it. What is more, if the Soviets had extracted oil more slowly in the 1970's and 1980's, then they would have had their oil crisis earlier. Gustafson (1989) points out that two "oil crises" occurred in the late 1970's and mid 1980's. However, neither of these so called crises caused reductions in oil production except in 1984 and neither caused economic problems. It was only the decline after 1988 that caused economic problems and was a true crisis. The fact remains, scarcity caused the Soviet oil production decline, just like it caused the US oil production decline. Whether it was Soviet mismanagement or not that caused the oil production decline, oil production did decline. There was oil scarcity. The economy had to adjust to less oil, which induced economic stagflation.

Another suspected cause for the production decline is a lack of investment. However, the level of investment in new and developmental drilling according to the CIA was increasing in the late 1970's and early 1980's. Gustafson says the number of exploration wells completed increased by 6.5% from 1965 to 1975, and capital spending on oil and gas exploration increased by 27% from 1980 to 1985 with even greater increases in investment in other oil and gas infrastructure needs. Certainly if any new giant oil fields existed, they could easily have been found. Furthermore, now that the Soviet Union has collapsed and the oil industry is more privatized, no one believes that substantial increases in investment now will create greater oil production then was attained in 1988 or even 1995. Some may claim that the reign of confusion in the industry during the Soviet break up may also have caused less investment during those years and even a loss of control of management. Again however, it is impossible to believe that a very few years of less than average incentives or management or investment can cause a 50%

reduction in the production of oil when before and after there is plenty of incentive, management and investment. If such a situation could happen then why did it not happen in Iran during its revolution in 1979. In Iran, the whole oil industry collapsed, but then a year later was back up to its former production level and under a new highly economically restrictive regime. Russia went from a restrictive economic regime to a more open free market regime. How can such openness create such a severe oil production decline?

Unlike other industries the demand for oil is strong due to high export prices. So unlike other industries, the Russians can sell as much oil as they produce, and if they could produce the oil in the archaic Soviet period, then there is no reason except scarcity for them not to be able to produce as much now. Also if the Soviet Union was able to transport its oil in and around Eastern Europe and export it, then there was certainly the export infrastructure to sell its oil after the collapse. The decline in production cannot be due to a decline in demand or a lack of export infrastructure alone.

However, the same arguments about the Soviet oil industry could also be heard in the US oil industry in the 1970's and 1980's: There was too much regulation, no incentive for new technology and so on. Yet the US and the USSR both had production declines, and the US endured this despite its comparatively freer market system and better property rights. Clearly since the US and the USSR had different economic systems, they should have produced oil in different patterns, but in fact they both show the exact same pattern of oil production.

Compare the Soviet oil experience to the US experience. In Figure 11.1 is the US oil production as a function of cumulative production. Dates for important events are added. Notice that the US encountered a steep decline in oil production even though prices were very high. It cannot be the case that low oil prices or a lack of investment caused US oil production to decline. During the 1970's and early 1980's investments in US oil production and new wells increased dramatically. Yet with higher oil prices and increased investment, oil production fell. During the 1970's and 1980's there was constant talk in the US about regulations or mismanagement causing the US production decline, yet even after deregulation in 1980 and freer markets, the US production continued to decline.

Figure 11.1. US Lower 48 Oil Production and Real Oil Price as a Function of Cumulative Production

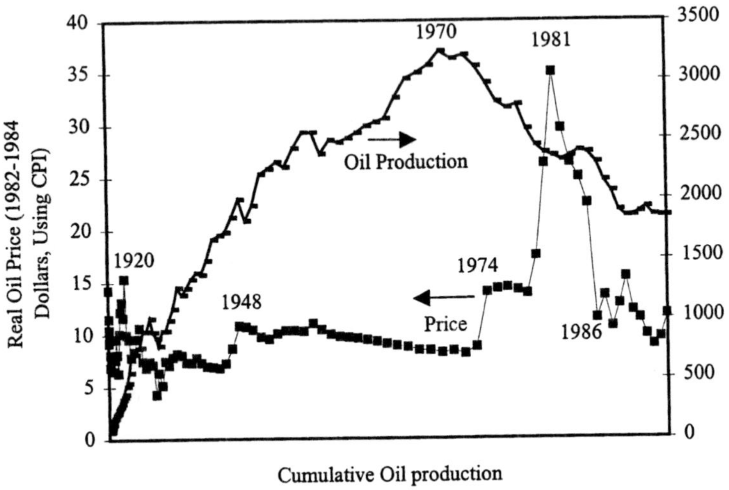

Now look at Figure 11.2. This is Soviet oil production as a function of cumulative production. It shows the exact same pattern as in the US. There is no reason to believe that the cause of this pattern is any different than the US. The USSR had incentive to find oil just like the US. It's just that Soviet natural resources are subject to the same laws of geology and supply that the US is. Incentives such as career advancement or even selling oil on the black market all combined to induce the Soviets to find and produce as much oil as possible and it did. The technology, the incentive, and the investment existed in the Soviet Union to find oil especially after production stagnated. The cause of the Soviet and Post Soviet oil decline then has to be for a similar reason as the US, namely scarcity. The US and Russia simply have less oil. Both are mature oil producing regions.

Figure 11.2. Soviet and Former Soviet Oil Production

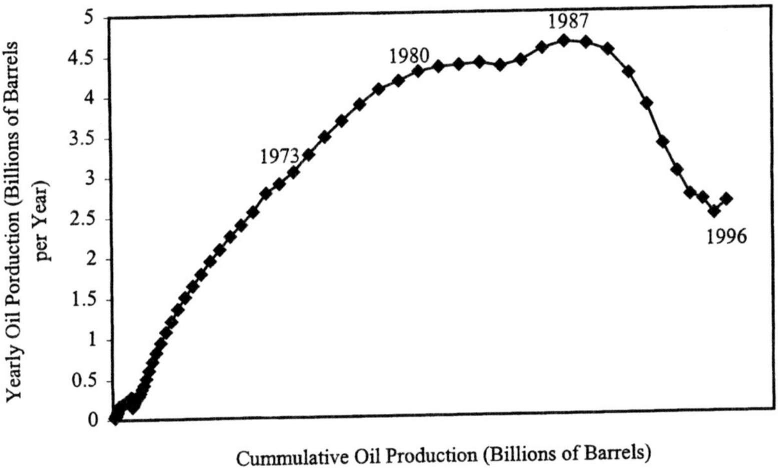

It has also been argued that if the Soviet's had invested more into Kazakstan's Tengiz oil field or the off shore oil fields of the Caspian Sea that they could have stopped the decline in production. However at a reserve-production ratio of 20 or so, the Tengiz can produce about 1 to 2 million barrels of oil a day and the Caspian may be able to double that. That is still not enough oil to compensate for the huge decline in Russian oil during the Soviet collapse. What is more, as in the case of Alaskan oil in the 1970's, these oil sources are remote and require huge new investments that will take time to get on line. Even if the Soviets had started in the 1970's to exploit these resources using Western technology, it would have required a much higher internal price for oil before they could have been accomplished. In other words, there would had to have been an oil price shock within the USSR in order to start developing these resources. Such a price shock would have pushed the Soviet Union into a collapse earlier than it did.

Russia, like the US, has to be considered a mature oil-producing region. Even if some Russian arctic or offshore regions have not yet been explored, neither has some arctic and offshore regions been explored in the US, yet no one believes that the US or Canada will have increasing oil production. Even the development of Prudhoe Bay itself did not stop US oil production decline after 1970. So what chance does a new giant oil field discovery in Russia have of increasing Russian oil production above 1988 levels, especially since Russia's oil production decline after 1988 was greater than the US's after 1970? The fact of the matter is the Soviets would have easily found a Prudhoe Bay size oil field if one existed. The technology to find such a large field is as old as the early 1900s. The Soviets, unlike the Americans in northern Alaska, were looking long and hard near the arctic coasts, and would have found something if it existed.

Since the collapse of the Soviet Union, Russian oil production did manage to increase slightly in 1997 undoubtedly due to new and better technology, new investment and better management. However in 1998 and the forecast for 1999, oil production is going down again probably due to lower prices. Still, slight changes in current production are incidental compared to the monumental decline suffered after 1988. Even if Russia were to find a Prudhoe Bay or two tomorrow, they could not climb back to their peak.

The fact of the matter is, a high price for oil implies scarcity. Scarcity caused the USSR to free market prices especially oil prices. The only way that the Soviets could do this was to free up their economy since energy and especially petroleum energy is the foundation of an economy. While we can certainly speculate that Russia's decline in oil production may have gone down faster than it needed to during the Soviet break up, it was bound to decline just like US production did. This created an oil price shock. Then just like the US had major economic recessions after the oil price shocks, so did Russia. The former communist countries both before and after their market transitions are still coping with higher oil prices.

THE CRUCIAL EVENTS FROM 1988 to 1992

In order to analyze what really happened in Eastern Europe and the Soviet Union during the crucial break up period, it would be necessary to have hard data on oil consumption and allocation. Unfortunately, official statistics during this period cannot be trusted since they were produced to pacify the public and quell the party managers. Still there are some interesting facts that can shed light on the period.

In the 1970's and 1980's the Soviet Union was trying to get Eastern Europe to cut its reliance on oil especially from the Soviet Union so that it could export more oil for foreign hard currency (CIA, 1985). Before 1989, the Soviets provided as much oil as Eastern Europe needed using a bartering system (FBIS, 1989). In reality the oil was given away at well below world prices (Balabanov and Deitz 1991). The Soviet Union was practically giving away its oil to Eastern Europe, Cuba and North Korea in order to hold up their economies and Soviet influence. As oil production waned, those exports declined and the subsidy was cut causing those economies to fall one by one alongside the Soviet economy. Already the Soviets had a severe credit crunch due to lower oil prices on their oil exports in 1986. The Soviets compensated by cutting foreign imports that year. Unfortunately exporting less lead to greater foreign currency problems, and so the Soviets started borrowing money to pay for imports. Eventually the Soviet and Eastern European economies were forced to cut oil consumption in order for the Soviet Union to be able to export oil for needed revenue to pay interest on foreign debts and buy essential Western technology.

Finally in 1989, the Soviets announced that trade to Eastern European Countries in CMEA (the Council for Mutual Economic Assistance) would be conducted in hard currencies and that oil exports to CEMA would be cut by 10% in 1990 (Balabanov and Deitz 1991). During that year and after, the Soviet Union undoubtedly cut oil exports, although exact figures are hard to come by. The Soviets probably went more quickly to selling their oil for hard currency than official accounts and undoubtedly cut exports to those who could not pay. Thus from 1989 to 1990, Eastern Europe was going through an oil crisis. With less virtually free oil from the Soviet Union, Eastern European countries had to adapt their economies to the new market. They had to become more oil efficient overnight. The only way was with freer markets. The Soviet Union could see that the only way to cut exports of free oil to Eastern Europe was by simply cutting off from the established trading system and let Eastern Europe sink or swim. In 1989 then, Gorbachev initiated total glasnost in Eastern Europe and stood back hoping it would simply stay in the Soviet sphere.

It was clear soon thereafter that with Eastern European economies in decline and with discontent rising, the Soviet Union had to decide between letting Eastern European countries leave the Soviet empire or controlling them militarily and giving them virtually free oil. Since Soviet oil was becoming a precious commodity, the Soviets had no choice but to let go of Eastern Europe. Without the free oil, the Eastern European economies went into a tailspin. Subsequent discontent pushed them toward democracy. The Soviet Union was left trying to simply keep NATO troops out of Eastern Europe but still letting the Eastern Europeans become democracies. CMEA was dismantled, (BW 1989a). No longer did the Soviet Union appear to have the will to push its communist ideal.

But the changes were not yet over. The Soviet Union itself was in decline. In the fall of 1989, with prices still fixed, a curious phenomenon occurred. Soviet wheat and other agriculture crops were rotting in the fields (BW 1989b). How could that have happened? We know that Soviet oil production was already declining and that oil prices inside the Soviet Union were still set well below world prices. Yet free markets were not introduced in the Soviet Union until early 1991. At first one might suspect rotting crops were due to a low artificial price that gave no incentive to harvest them. However, there were in fact black markets all over the country at that time. Harvested crops could have easily been sold if they could simply be transported to a black market. Ellman (1992) suggests that it was changes in administrations with continuous reorganization of the agricultural complex that caused it. However, again, black markets give incentives that can overcome poor government planning.

The problem was more likely a lack of petrol. Gasoline or oil sold on international markets would have provided a tempting source of hard currency reserves for the government in order to import needed food and technology. Undoubtedly much Soviet oil was being sold by the Soviet authorities or by black marketers on spot world markets. It would have been especially tempting for the government itself to sell on the world markets in order to pay interest payments on loans. With more oil sold abroad, there would be oil shortages in agriculture. And since agriculture is a highly petroleum-intensive industry, it is no wonder crops were left to rot on farms. No one had petrol to harvest the crops and transport them to market.

Eventually in 1991, all prices were allowed to go to market levels, but it takes time for an economy to adjust. So the economy fell into a deeper and deeper recession. Note too that even though prices were free, oil prices were kept low and the economy was not allowed to allocate the oil to its most productive uses, i.e. to its highest bidder. This meant that oil was sold on world markets, creating shortages at home. Inefficient factories and producers were allowed to continue to receive wages even though they produced nothing that the economy demanded. It was this lack of flexibility in Eastern Europe and the Soviet Union that helped their economies spiral downward. However, the initiating event was an oil crisis. Even in 1997, after ten years of *perestroika* and seven years of free market prices, many factories and business were still being supported artificially by the state. Many firms were still receiving low-priced oil. Former Soviet oil companies have been required to sell fuel at prices below world market prices. This has prolonged the inefficiencies of the Eastern economies. However, even with freer market prices, oil production continues to decline, with only the slight increase in 1997.

During the transition period then, not only were Soviet industries having a hard time coping with higher oil prices, but also the economy as a whole had less oil to export. Low oil exports contributed to a balance of payment and international debt problem that Russia still faces today. The internal oil price shock, bankrupted Soviet industry. All of these crushing economic problems were brought about mainly by a shortage of oil. Eastern Europe and the Soviet Union really had only one course of action: free the markets. However it was impossible to free their markets without also opening up their political systems too-since the state owned everything.

CONCLUSION

To suggest that the break of the Soviet Union caused the decline in oil production would be like saying the energy crises in the US caused the US oil production decline. Incentive to produce oil, even if on the black market, always existed. Most likely within the conflicting statistics, the period of the break of the Soviet Union was an oil crisis period. We know that the West went through two severe oil crises, which the Soviet Union and Eastern Europe never felt. It was only a matter of time for them finally to have their own oil crisis. They simply insulated themselves from it for a while. The fact remains that former Soviet oil production by 1992 was 20% less than Soviet and Eastern Europe oil consumption in 1988. Furthermore, incredibly, oil production in the Soviet Union declined *before* consumption, which makes a cause and effect scenario of an economic crisis causing an oil production decline impossible. The more likely cause and effect scenario is an oil crisis caused by scarcity leading to an economic down turn and eventual consumption reduction. This means that these economies were forced into an oil crisis. They had to reduce oil use. However, with vastly higher oil prices, actual former Soviet and Eastern European oil consumption declined by about 50% from 1988 to 1995.

When the West cut its oil consumption in 1974 and 1981, there were two major recessions. However, neither of those recessions and energy efficiency periods saw a 50% reduction in oil usage in the West. Considering the West's experience, there would be no way for the planned economies of Eastern Europe and the former Soviet Union not to have had even bigger recessions during the period 1988 to 1995-when they saw an even bigger reduction of their oil supply-than the West had in the 1970's and 1980's. Logically, this means an oil crisis pushed these planned economies towards a more efficient system, namely free markets and democracy.

REFERENCES

Balabanov, Todor and Raimund Dietz (1991). " Eastern and East West Energy Prospects" Dismantling the Command Economy in Eastern Europe, Peter Havlik Ed. Westview Press, Boulder CO, pp. 125, 137.

Boyd, Roy and Tony Caporale, (1996). "Scarcity, Resource Price Uncertainty and Economic Growth" Land Economics, Volume 72, Number 3, August pp. 326 -335.

BP (1996) BP Statistics Review of World Energy, BP oil company, London.

Brown, Thomas G. (1992). "The Soviet Union as a Great Power: The need for Reform," The American Economist, Vol. 36, No 1, spring, pp. 77 -84.

BW, Business Week (1989a) "A Talk with Nikolai Ryzhkov, Economics Czar," June 5, pp. 62-63.

__________ (1989b) "A Blight on the Fruited Plain," June 5, p. 57.

Central Intelligence Agency (1985). USSR Energy Atlas, Washington DC pp. 12 -28.

__________, Directorate of Intelligence (1988) Revisiting Soviet Economic Performance under Glasnost: Implications for CIA estimates, SOV 88-10068, Washington DC CIA September.

Dallin, Alexander (1992). "Causes of the Collapse of the USSR," Post-Soviet Affairs, Vol. 8, no. 4, pp. 279 - 302.

Economic Report of the President (various years).

Ellman, Micheal (1986). "The Macro-Economic Situation in the USSR – Retrospect and Prospect" Soviet Studies, Vol. 38, No. 4 pp.530 - 542.

__________, and Vladimir Kontorovich (1992) "Over View," in The Disintegration of the Soviet Economic System, Micheal Ellman and Vladimir Kontorovich Eds. Routledge N.Y., pp. 1- 39.

Energy Information Administration (Various years) Energy Yearbook.

FBIS,US Foreign Broadcasting Information Service (1990) Soviet Union Economy 18 December 1990 p. 71.

__________ (1989) Soviet Union Economy 12 May 1989 p. 38.

Gisser, Micha and Thomas H. Goodwin, (1986). "Crude oil and the Macroeconomy: Tests of Some Popular Notions," Journal of Money and Banking, Volume 18, pp. 120-129.

Gustafson, Thane (1989). Crisis Amidst Plenty: The Politics of Soviet Energy under Brezhnev and Gorbachev, Princeton University Press, Princeton, NJ.

Kafouros, Wassily (1996). "Towards a Dynamic Two Sector Model of Soviet Economic Decline," Review of Radical Political Economics, Vol. 28, No. 3, pp. 35 - 49.

Kotz, David M. and Fred Weir (1997). Revolution From Above, The Demise of the Soviet System, Routledge, London and New York pp.34.

Krueger, Gary (1993). "Goaszakazy and the Soviet Economic Collapse" Comparative Economic Studies, Volume 15, Number 3, fall pp. 1 - 18.

Osband, Kent (1992). "Economic Crisis in a Shortage Economy," Journal of Political Economy, Vol. 100, No. 4, pp. 673 - 690.

Reinsh, Anthony E., Igor Lavrovsky and Jennifer I Considine (1992) Oil in the Former Soviet Union: Historical Perspective Long-term outlook, Canadian Energy Research Institute, Alberta Canada.

Time (1975) "International Issues: Henry Kissinger's views on possible US military intervention against the oil producers," Vol. 105, January 20, pp. 34-35.

Index

1549, 123
1560, 121
1660, 121
1691, 123
1700th Century, 121, 124, 138
1709, 111, 164
1714, 123
1760, 124
18th Century, 111, 118, 121, 122, 134, 164, 175
1803, 123
1902, 20, 59
1913, 160
1920s, 77
1930, 20, 59
1940s, 136
1950s, 83, 136
1960s, 83, 136, 209
1960, 203, 204
1969, 97
1970s, 2, 10, 15, 47, 71, 72, 89, 99, 105
1970, 5, 53, 72, 97, 98, 214
1973, 10, 47, 49, 51, 53, 54, 55, 70, 72, 73, 74, 89, 92, 96, 98, 101, 102, 105, 107, 129, 157, 158, 198, 204, 205
1975, 203, 204, 205, 208, 210
1979, 10, 47, 48, 49, 51, 54, 99, 106, 107, 175, 204, 211
1980s, 10, 70, 89, 99, 201, 202, 203, 205, 206, 209, 210, 211, 215, 219
1980, 47, 49, 50, 53, 55, 70, 72, 91, 203, 204, 208, 210, 212
1982, 72, 73, 89, 90, 91, 198, 204
1983, 72, 101, 102, 131, 137, 138
1985, 199, 200, 204, 210
1986, 215
1988, 197, 198, 203, 204, 207, 208, 210, 214, 215, 218, 219
1989, 198, 202, 206, 215, 216
1990s, 201, 206, 208, 209, 210
1990, 49, 50, 51, 53, 55, 215, 216
1992, 197, 203, 204, 207, 215, 218
1993, 72, 198
1995, 199, 206, 210, 218, 219
1997, 83, 214, 217
1998, 92, 100, 175, 214
2030, 95, 97, 100
2037, 100, 107
22nd Century, 95, 97

A

Adelman, M.A., 48, 55, 70, 72, 73, 86, 92, 109
air travel, 182
airplanes, 133, 134, 167
Alaska, 130, 213, 214
    Prudhoe Bay, 25, 146, 171, 214
Algeria, 83
Al-Jarri, Abdulrahman S., 43, 46, 55
Al-Sultan, A.M., 70, 92
American Petroleum Institute (API), 47, 48, 51, 53
API. *See* American Petroleum Institute
Apostolakis, B.E., 187, 194
Arab-Israeli War. *See* Six Day War
Arab oil embargo, 105, 141
ARAMCO, 82, 96
arctic, 146, 214
area grade. *See* grade, energy
Ashton, Thomas Sutheliffe, 122, 123, 129

assembly line, 160-2
astronaut, 157
Atkinson, G.D., 195
Atsegbua, L., 83, 92
auto correlation, 50, 51, 54
Axelrod, R., 88, 92

B

Balabanov, Todor, 203, 215, 219
Barnes, Harry Elmer, 129
Barnett, Harold J., 3, 4, 5, 15, 18, 38, 158
barrel of oil equivalent (BOE), 148, 150
batteries, 113, 134, 153, 154, 163, 181, 182
Behrens, W.W., III, 158
Berg, Doug, 110
berries, wild, 117-122
biomass, 151, 152, 169, 170, 181, 195
Black Box, 41, 42, 44
BOE. *See* barrel of oil equivalent
bonanza, 17, 43
Boyd, Roy, 202, 219
Breusch-Godfrey test, 65
Bronowski, Jacob, 164, 176
Brown, G.M., 182, 195
Brown, Thomas G., 201, 219

C

"C" shaped production function, 187, 188. *See also* substitutes
Campbell, Colin J., 55
Cameron, Rondo, 111, 129
Canada, 150, 214
capital. *See also* investment
  extraction, 114
  information. *See* Robinson Crusoe Model
  input, 132-34, 137-38, 148, 150, 155, 161, 162, 168, 171-173
  investment, 77, 78, 81, 85, 11, 122, 128, 137, 171, 177-9. *See also* Soviet Union.
capital-energy ratio. *See* substitutes
capital-labor ratio. *See* substitutes
capital-land ratio. *See* substitutes
capital-pollution ratio. *See* substitutes
Caporale, Tony, 202, 219
car. *See* electric vehicle, gasoline, propane automobile, and solar
cartel, 2, 969, 70, 72, 73, 74, 92, 93. *See also* cheating, collusion, game theory model, incentive to cheat and OPEC
Caspian Sea, 213
Central Intelligence Agency (CIA), 203, 204, 210, 219
Chapman, P.F., 58, 68
cheating, 88, 89, 90, 91, 93. *See also* cartel, collusion, game theory model, incentive to cheat, and OPEC
CIA. *See* Central Intelligence Agency
Cipolla, Carlo M., 129
Cleveland, Cutler J., 31, 38, 44, 55, 58, 68, 98, 110
CMEA. *See* Council for Mutual Economic Assistance
coal, 10, 23, 54. *See also* England, resources, and technology
  as alternative, 150, 162-75
  conversion, 150, 151, 154, 157
  entropy, 134-139
  grades, 111, 113, 115, 117, 118, 128
  history of use, 117, 118, 121-24, 128
  Industrial Revolution, 121-125
  as part of energy utilization chain (EUC), 145, 148, 150, 151, 154, 157
Cobb-Douglas Production Function, 177, 182, 184
collusion, 70, 72, 87. *See also* cartel, cheating, game theory model, incentive to cheat, and OPEC
combustion engine
  external, 151, 166-68
  internal, 151, 156, 166-68, 172
command (planned) economy, 197, 201, 203, 205, 206, 208, 209, 219. *See also* Soviet Union
Communism, 14, 138, 197, 202, 207-09, 214, 216. *See also* Soviet Union
conservation, 39, 104, 109, 138, 143, 145, 152, 153, 154, 156, 204. *See also* energy utility chain (EUC), technology *and* Soviet Union
Considine, Jennifer I., 220
consumer price index (CPI), 47
consumption of energy, 55, 96, 99, 100, 101, 106, 107, 143, 145, 170, 194. *See also* Soviet Union, technology and Robinson Crusoe Model
  on Hubbert curve, 96, 101, 106, 107
  trends, 96, 99, 100, 101, 106, 107, 218
conversion of energy. *See* energy
corn economy model. *See* Robinson Crusoe Model
Council for Mutual Economic Assistance (CMEA), 215, 216. *See also* Eastern Europe and Soviet Union
CPI. *See* consumer price index

Crausby, Richard, 123
Cremer, J., 70, 92
crude oil, 38, 51, 53, 110, 146, 148, 149, 219
Crusoe, Robinson. *See* Robinson Crusoe Model
Cuff, D.J., 149, 158

D

Dahl, C., 70, 72, 92
Dallin, Alexander, 200, 202, 219
Danielsen, A.K., 70, 93
Darby, Abraham, 111, 123, 164
Dasgupta, P.S., 177, 178, 183, 194
Department of Energy (DOE), 89, 93, 109
depletion effect
  exploration, 6, 7, 67, 68
  reserves, 5, 110
depreciation, 185, 186. *See also* entropic decay
diesel, 151, 152, 155, 167
Dietz, Raimund, 219
DeVries, John, 129
DOE. *See* Department of Energy
dominant player. *See* price leader
Dubourg, W.R., 195
Duncan, D.C., 136, 140

E

East Texas, 20, 25, 59, 60, 62, 77, 132, 171
Eastern Europe, 198, 199, 201, 202-07, 211, 215, 216, 218, 219
economic epoch, 111, 112, 118, 128, 129
economic growth, 34, 92, 100, 111, 112, 117, 121, 140, 141, 142, 143, 158, 159, 164, 176, 178, 195. *See also* limits to growth and Soviet Union
economic stagnation, 111. *See also* Soviet Union
economies of scale, 11, 70, 81, 122, 124, 152, 155, 159-70, 173, 175, 176
Edwards, John D., 31, 38, 98, 110
EIA. *See* Energy Information Administration
elasticity of
  demand, 70
  substitution, 177, 178, 179, 182, 183, 184, 186-70
  supply, 7, 45, 53, 56, 99. *See also* "C" shaped production function
electric power, 10, 116-17, 125, 128, 134, 144-45, 153-55, 163, 167-69, 172-75, 182, 193. *See also* energy utilization chain (EUC) and technology
electric vehicle, 113, 125, 130, 134, 145, 151, 153, 154, 163, 174, 175
electricity EUC. *See* energy utilization chain (EUC)
Ellman, Michael, 199, 200, 219
embargo, oil. *See* Arab oil embargo
energy. *See also*, coal, conservation, consumption of energy, diesel, electric power, energy –labor ratio, energy utilization chain (EUC), ethanol, exhaustible resources, fuel, gasoline, geothermal, grade, heavy oil, hydropower, methanol, natural gas, non-renewable resources, nuclear, oil, production, propane, reserves, resource, scarcity, statistical analysis, steam, substitutes, solar, technology, *and* uranium
  consumption link, 143, 145
  conversion, 116, 125, 136, 140, 143, 145, 148, 150, 151, 167, 169, 170-74
  conversion link, 143, 145. *See also* grade, energy
  service link, 143, 145
  shortages. *See* shortage, energy
  source link, 143, 145
  theory of value, 125
Energy Information Administration (EIA), 10, 45, 47, 48, 51, 53, 56, 93, 95, 96, 97, 99, 100, 101, 103, 105, 106, 107, 109, 110, 175, 176, 204, 219
energy utilization chain (EUC), 12, 143-48, 151-54
  electricity, 145, 153-55
  lifestyles, 154-56
  oil conservation, 145, 152, 153-54, 156
  oil substitutes, 145-48, 151-52, 157
  oil use, 144-46, 153, 175
energy-capital ratio. *See* substitutes
energy-labor ratio. *See* substitutes
England, 111, 118, 121, 124, 125, 134, 138, 169, 175. *See also* coal, Darby, and Industrial Revolution
entropic decay, 185, 186, 189. *See also* depreciation
entropy, 135, 136, 138, 177, 179, 180, 184-89, 193, 194
  subsidy, 11, 12, 37-39, 131, 135, 136-39, 148-50, 158, 171, 172, 176
environment, 38, 56, 68, 110, 151, 195

awareness, 159
hazards, 116, 128
costs, 139
problems, 139, 167, 169
epoch, economic, 111, 112, 118, 128, 129
epochal innovation, 111, 128. *See also* economic epoch and technology
essential resource, *See* resource
estimated results of world oil supply. *See* statistical analysis
ethanol, 117, 147, 151, 169, 170. *See also* energy, fuel, methanol, and resource
EUC (see energy utilization chain)
exhaustible resource, 3, 15, 38, 56, 68, 93, 140, 158, 194, 195. *See also* non-renewable resource and resource
exhaustion, 3, 15, 18, 179, 194
exploration, 1, 8, 10, 17, 20, 39, 42-45, 56, 57, 62, 75, 76, 77, 80-82, 85, 86, 110. *See also* energy, Mayflower Problem, oil, resources, Robinson Crusoe Model, scarcity and technology
costs, 17-19, 23, 33, 39, 57, 58, 110
East Texas, 59, 62
efforts in OPEC countries, 9, 69-72, 75, 83, 86, 90, 91, 96, 109
efforts in non-OPEC countries, 9, 71, 96
efforts in Soviet Union, 210
efforts in the US, 1090
Mayflower Problem, 57, 58, 62
mineral economy model, 20, 22, 24-30, 33, 34
Monte Carlo Simulation, 57-59, 62, 65, 67
technology, 144, 146
external combustion engine. *See* combustion engine
extraction. *See also* capital, energy, oil, resource, mineral economy model, technology
cost of , 3, 5, 17, 18, 19, 25, 26, 30, 33, 34, 58, 86
quantities, 17, 38
resource, 2, 5, 6, 17, 53, 96, 99, 103, 114
technology, 144, 146
Exxon, 83
Ezzati, A., 70, 93

F

factory, 160, 162, 165, 199. *See also* investment
FBIS. *See* Foreign Broadcasting Information Services
field state grade. *See* grade, energy
Fiegenbaum, A., 71, 76, 80, 93
Fishburn, P.C., 69, 71, 80, 75, 93
Fisher, A., 58, 68
fixed stock paradigm, 142, 177, 178, 195
Ford, Henry, 160-62
forecasting. *See* statistical analysis
Foreign Broadcasting Information Services (FBIS), 219
Friedman, M., 69, 71, 75, 76, 93
fuel. *See also*, coal, conservation, consumption of energy, diesel, electric power, energy, energy –labor ratio, energy utilization chain (EUC), ethanol, exhaustible resources, gasoline, geothermal, grade, heavy oil, hydropower, methanol, natural gas, non-renewable resources, nuclear, oil, production, propane, reserves, resource, scarcity, shortage, statistical analysis, steam, substitutes, solar, technology, *and* uranium
cell, 154, 174
liquid, 113, 115-17, 126, 133, 134, 147-50, 154, 163, 166, 169, 170-72
solid, 113, 115-117, 133, 147-49, 151, 163, 166, 169, 171
synthetic. *See* synthetic fuels
fuel-efficiency, 153
capital, 205
in transportation, 205

G

gambler, 20, 21, 24, 26, 29, 33, 60, 61, 62, 71, 77. *See also* game theory model, Mayflower, Monte Carlo simulation, Robinson Crusoe Model, OPEC, and risk-loving behavior
game, wild, 117-122
game theory model, 57-68, 69, 73-75, 88, 89, 91, 93. *See also* cartel, cheating, collusion, incentive to cheat, Monte Carlo simulation and OPEC
disutility of losses, 84
gas-to-liquids (GTL), 147, 148, 170, 171. *See also* technology
gasoline, 85, 125, 133, 134, 143, 145, 148, 152, 153, 163, 167, 170, 171, 217. *See also* technology
cars, 114, 125, 130, 134, 143, 145, 151-154, 160, 163, 167, 168, 175, 184, 190

trucks, 11, 128, 151, 153, 155, 167, 168, 187
GDP. *See* gross domestic product
geological characteristics, 78
geophysical, 43
geothermal, 154
Georgescu-Roegen, Nicholas, 1, 4, 17, 38, 43, 56, 179
Ghwar, Saudia Arabia, 25
Gisser, Micha, 202, 219
glasnost, 22, 201, 202, 216, 219. *See also* Soviet Union
Goldemburg, J., 158
Goodwin, Thomas H., 202, 219
Gorbachev, Mikhail, 200, 202, 216, 219. *See also* Soviet Union
Gordon, R.B., 147
grade, energy, 111-117, 132. *See also* energy
area, 113, 114, 117, 121, 122, 124, 128, 132, 133, 134, 147, 148, 151, 161, 162, 164, 165, 166, 169, 170, 171
characteristic, 116, 117, 119, 132, 148, 166
high, 11, 111-15, 117-21, 124, 125, 127-29, 131, 134-40, 147, 148, 150, 158, 161, 162, 164, 166, 167, 170, 171, 174, 175, 192
low, 11, 112, 114, 116, 117, 124, 125, 127-29, 134-40, 147, 148, 151, 153, 155, 158, 161-64, 168-74, 192
state, 113-117, 132-134, 147-149, 151, 153, 161, 163, 164, 170, 171, 191
state, field, 116-117, 133, 134, 163, 168, 169, 173: electric, 116, 117; magnetic, 116, 117; pressure, 116, 117; radiative, 117, 168, 172
state, gas, 115, 117, 167
state, liquid, 115, 117, 126, 133, 149, 163, 166
state, solid 11-117, 149, 164, 169, 171
volume, 113, 114, 115, 117, 128, 132, 133, 147, 148, 151, 161-64, 166-68, 170, 174
weight, 113, 117, 122, 128, 132, 133, 147, 148, 151, 153, 155, 158, 161-64, 168, 170
Graham, S., 131, 140
grain, 117-122, 124, 169, 170
Griffin, J.M., 70, 72, 74, 93
gross domestic product (GDP), 83, 86, 125, 143, 144, 146, 152, 155-57, 175, 180, 194, 202
growth, economic, 34, 92, 100, 111, 112, 117, 121, 140, 141, 142, 143, 158, 159, 164, 176, 178, 195. *See also* limits to growth and Soviet Union
Gruen, William, 44, 56
GTL. *See* gas-to-liquids
Gulen, S.G., 70, 72, 93
Gulf War, 49
Gurney, J., 83, 93
Gustafson, Thane, 198, 210, 219

H

Hall, Darwin C., 19, 38
Hall, Jane V., 19, 38
hardtack economy. *See* Robinson Crusoe Model
Hartwick, J.M., 13, 15, 177, 178, 183, 194
Heal, G.M., 177, 178, 183, 194
heavy oil, 139, 146, 149, 150, 171, 172
Hotelling, Harold, 1, 3, 5, 7, 15, 17-19, 32, 34, 35, 37-39, 49, 56, 68, 140
Hubbert, M. King, 1, 4, 6, 7, 15, 38, 55, 56
Hubbert curve, 1, 4, 6, 7, 8, 10, 30-31, 32, 35, 37, 41, 42, 44-46, 52, 54, 55, 56, 59, 65. *See also* production and scarcity
life cycle, 30, 31
limit, 7
non-time dependent, 41, 42, 46
quadratic, 7, 41, 44, 45, 54, 55
hydrogen, 117, 174
hydropower, 116, 154, 181

I

incentive to cheat, 70, 74. *See also* cartel, cheating, collusion, game theory model and OPEC
Industrial Revolution, 111, 112, 118, 124, 129, 160, 164-69, 175. *See also* coal, Darby, and England
industrialization. *See* Soviet Union
infrastructure. *See* investment
injection wells. *See* oilwells
innovation, technological. *See* epochal innovation, technology, *and* Soviet Union
INOC. *See* Iran National Oil Company
intergenerational equity, 15, 194, 195
internal combustion engine. *See* combustion engine
International Oil Company (IOC), 82

investment. *See also* capital
factory and infrastructure, 160, 166, 171, 201
oil development, 74, 75, 77, 81, 82, 85, 87-91, 104, 109, 208, 210-214
Soviet Union oil, 208, 210-214
IOC. *See* International Oil Company
Iran, 10, 82, 96, 104, 106, 107, 211
National Oil Company (INOC), 83, 86
Revolution, 106, 211
Iraq, 104
Isfahani, D.S., 70, 92

J

Johanson, Thomas B., 176
Johany, A., 70, 93
Joiner, C.M. "Dad", 77

K

Kang, H., 182, 195
Kaufmann, Robert K., 31, 38, 48, 55, 56, 98, 110
Katell, S., 131, 140
Kazakstan, 217
Kissinger, Henry, 220
Kochenberger, G. A., 69, 71, 75, 80, 93
Kontorovich, Vladimir, 199, 219
Koopmans, T.C., 194
Kotz, David M., 199, 202, 219
Kuznet, Simon, 111
Kuwait, 104

L

labor-capital ratio. *See* substitutes
labor-energy ratio. *See* substitutes
labor-land ratio. *See* substitutes
Laherrere, Jean H., 42, 55, 99, 110
land-capital ratio. *See* substitutes
land-labor ratio. *See* substitutes
large independent mobile machinery (LIMM), 143-47, 153, 154, 160, 167, 174, 175, 182
Lavrovsky, Igor, 220
law of entropy. *See* entropy
lifestyles EUC. *See* energy utilization chain (EUC)
limits
to growth, 142, 158, 175, 176, 177, 195
to substitution, 179
LIMM. *See* large independent mobile machinery
Loderer, C., 70, 72, 89, 93
long-run, 19, 24, 27-30, 32, 37, 52, 103, 187, 188, 192, 209
lottery, 63, 64, 75
Lovejoy, D., 178, 195
low discount theory, 70
lower 48, the, 38, 41, 42, 45, 47, 51, 53, 54, 56, 57, 59, 69, 97, 98, 110. *See also* Alaska and United States
low-grade energy. *See* grades, energy
Lynch, Michael C., 97, 98, 110

M

Maine, 60
maintenance, 184-86. *See also* "C" shaped production function and substitutes
costs, 125, 172
replacement ratio, 187-89
Malthusian scarcity, 3, 19, 23. *See* Ricardian scarcity and scarcity
Markowitz, H., 71, 93
maximum usefulness line, 190
Mayflower Problem, 1, 17, 19, 20, 22, 23, 30, 43, 58. *See also* exploration, Hubbert curve, oil and supply
McCabe, Peter, 3, 39
Meadows, D.H., 158
Meadows, D.L., 158
Menard, H.W., 8, 9, 15, 20, 38, 59-61, 68
methanol, 117, 126, 130, 147, 148, 151, 154, 171, 174. *See also* energy, ethanol, fuel, and resources
Mexico, 9, 19, 72, 75, 82, 87, 96, 104
Middle East, 86, 114, 150, 206. *See also* Ghwar, Iran, Iraq, OPEC, Saudi Arabia, and United Arab Emirates (UAE)
Mikesell, R.F., 141-158, 175, 176, 178, 195
mineral economy model, 17, 22, 23, 24, 30, 31, 32, 56, 68, 110. *See also* Monte Carlo simulation, reserves, Robinson Crusoe Model, and supply
monopoly, 73, 82, 83, 85, 86, 87, 88, 91, 104. *See also* OPEC and Saudi Arabia
price leadership, 189, 191
Monte Carlo Simulation, 20, 21, 26, 29, 33, 57-64. *See also* exploration, mineral economy model, oil and supply
Montesi, Ronald, 44, 56
Moran, T., 70, 93

Moroney, John, 55, 98, 110
Morse, Chandler, 3, 4, 5, 15, 18, 38, 158

N

natural gas, 2, 10, 23, 54, 55, 56, 72, 93, 106, 113-117, 124-27, 130, 133, 139, 140, 145, 147, 148, 154, 162-64, 167-75, 203, 210. *See also* energy, fuel, oil, oilwells, supply and resources
  liquid, 113, 115-17, 126, 133, 134, 147-50, 154, 163, 166, 169-72
  solid, 113, 115-17, 133, 147-49, 151, 163, 166, 169, 171
natural resources. *See* exhaustible resources, non-renewable resource and resource
negotiation, 70, 74, 83, 85, 89
net present value (NPV), 71, 75-88
Neilson, W. S., 73, 93
Nigeria National Petroleum Corporation (NNPC), 83
NNPC. *See* Nigerian National Petroleum Corporation
non-renewable resource, 7, 179, 180, 182. *See also* exhaustible resource and resource
Nordhaus, W.D., 194
Norgaard, R.B., 1, 4, 17, 19, 39, 43, 56, 58, 68
Norway, 72
NPV. *See* net present value
nuclear energy, 116, 117, 127, 163, 168, 172, 191. *See also* energy, fuel, resource, supply, and uranium
  as alternative to oil, 10, 144, 173
  fuel cycle, 168
  fusion, 128, 144, 156, 172
  hazards, 168-9, 173
  power, 127, 128, 154, 168, 173, 191-92
nuts, wild, 117-122

O

OECD. *See* Organization for Economic Cooperation and Development
oil. *See also*, coal, conservation, consumption of energy, diesel, electric power, energy, energy –labor ratio, energy utilization chain (EUC), ethanol, exhaustible resources, fuel, gasoline, geothermal, grade, heavy oil, hydropower, investment, methanol, natural gas, non-renewable resource, nuclear, production, reserves, resource, scarcity, statistical analysis, steam, substitutes, solar, technology, *and* uranium
  alternatives. *See* resource
  crisis, 2, 14, 15, 99, 109, 152, 197-99, 203-08, 210, 216-219
  crude oil, 38, 51, 53, 110, 146, 148, 149, 219
  development, 82, 88
  embargo. *See* Arab oil embargo
  exploration game. *See* Monte Carlo Simulation
  heavy oil, 139, 146, 149, 150, 171, 172
  price shock, 1, 7, 10-11, 37, 38, 47, 49, 56, 70, 92, 99, 105-110, 131, 157, 159, 172. *See also* Soviet Union
  reserves, 25, 41, 42, 51, 72, 78, 87, 88, 93, 99, 104, 146, 149, 150
  shale oil, 12, 117, 131, 132, 136-40, 145, 146, 148-50, 157, 171
  shortages. *See* shortage, energy
Oil and Gas Journal, 12, 93, 130
oil conservation EUC. *See* energy utilization chain (EUC)
oil shale. *See* oil, shale
oil substitutes EUC. *See* energy utilization chain (EUC)
oil use EUC. *See* energy utilization chain (EUC)
oilfields, 25, 58, 63, 69, 71, 80, 81, 86, 103, 104, 132, 205, 209, 213
  giant, 20, 59
  supergiant, 8, 20, 59, 77
oilwells, 20, 22, 47, 48, 59, 60, 62-64, 78, 84, 85, 103, 104, 125, 150, 210, 211. *See also* oil, fuel, resources and supply
  injection, 78
  natural gas, 170
  wellhead, 47, 48, 125
OPEC, 2, 9, 10, 69-75, 82, 86-93, 96, 104-07, 109, 110. *See also* cartel, cheating, collusion, game theory model, incentive to cheat, production, and reserves
opportunity cost, 66, 142, 170, 177
opportunity cost paradigm, 142, 177, 178, 195
Orina River, Venezuela, 150
Osband, Kent, 199, 200, 219
Othmer, Donald F., 130

P

Paddock, James L., 25, 39
Page, Talbot, 22, 39
peak, 6, 31-42, 95-100, 109, 214. *See also* Hubbert curve and supply
Pearce, D.W., 177, 178, 195
Pearse, Peter, 18, 39
PEMEX, 82, 96
perestroika, 200, 217. *See also* Soviet Union
Persian Gulf, 20
Pesaran, M.H., 98, 110
petroleum. *See* oil
physics, 1, 2, 13, 43, 112, 177
Piceance Creek, Colorado, 149, 171. *See also* oil, shale
Pilgrims, 19, 20, 58. *See also* Mayflower Problem
Pindyck, R.S., 70, 93
planned economy. *See* command economy
power plant, 172, 173, 191, 193
Poland, 205
pollution, 139, 156, 167-69, 190, 191, 194. *See also* environment
PPI. *See* producer price index
price, 1, 3, 5, 7, 10-12, 17-19, 34-59, 68-74, 78, 85-99, 102, 105-110, 121, 125, 126, 129, 131, 132, 135-142, 146, 148, 151-159, 171-173, 182, 183, 198-219. *See also* Soviet Union, statistical analysis and supply
  ceiling, 89
  leader, 69, 73, 74, 87, 89, 91, 99
  leadership, 73, 87, 89, 91
  monopoly power, 9, 73, 89, 91
price shock, oil, 1, 7, 10-11, 37, 38, 47, 49, 56, 70, 92, 99, 105-110, 131, 157, 159, 172. *See also* Soviet Union
probability, 8, 20-22, 26-30, 32, 58-62, 65-67, 76-82. *See also* exploration, game theory model and OPEC
  coin, 26
  subjective, 21, 22, 26, 27, 28, 59, 61, 76, 78, 79, 81, 87
producer price index (PPI), 47, 48
production, 2, 5, 8, 9, 15, 18, 30, 31, 112, 119, 122-124, 127-129, 132, 134-138, 141, 144, 146, 149, 151, 160-162, 164, 166, 170, 172, 175, 177, 178, 180-191, 194
  capacity, 69-75, 88, 90, 91, 93, 106, 159, 177. *See also* "C" shaped production function
  Hubbert curve, 34-39,
  OPEC, 70-75, 78, 82, 84-93
productivity, 28, 112, 128, 129, 159-161, 167, 175, 178, 179, 193
  in US, 41-49, 52-58
  for US and Russia, 197-199, 202-218
profit, 8, 9, 33, 34, 37, 47, 48, 70, 73, 74, 77, 78, 82, 83, 86, 90, 91
propane automobiles, 133
property rights, 8, 9, 78, 90, 211
proven reserves, 48, 51, 95, 96, 102-104, 150. *See also* oil and reserves
Prudhoe Bay, Alaska, 25, 146, 171, 214
Pyramids of Egypt, 143

Q

quota, 70, 73, 74, 89-91

R

railroads, 166. *See also* Texas Railroad Commission (TRC)
Raistrick, A., 12, 129
Randers, J., 158,
random search, 8, 9, 59, 60. *See also* exploration, game theory model, Monte Carlo simulation and roulette
Ramsey, J.B., 71, 85, 93
rate of technical substitution, 183, 186, 187
Reagan, Ronald, 201
recession, 144, 153, 176, 198, 204-06, 214, 217, 218
Reddy, Amulza, 176
Reinsch, Anthony E., 209, 220
regulation, 47, 48, 49, 55, 83, 98, 99, 211, 212. *See also* price
research technologies, 144
reserves, 6, 146, 148-151, 213, 217. *See also* heavy oil, shale oil, oil, statistical analysis and shortage
  mineral economy, 17, 18, 23, 25, 26, 30-35
  OPEC, 72-75, 78, 86-88, 90, 93
  reserves-to-production, 70-90, 90-110
  ten-to-one ratio, 95, 96, 103-109
  ultimately recoverable reserves (URR), 42, 46, 96, 98-102, 105-108
  US, 41, 42, 46-48, 51, 56
  world, 95-97, 99, 101-110
resource, 2, 3, 6, 7, 11, 13, 18-20, 22, 29, 32-34, 36-39, 43, 55, 56, 68, 110, 114,

124, 125, 128, 142, 144, 158, 159, 174, 176-181, 191, 195, 204, 219. *See also* exhaustible resource, extraction, non-renewable resource and technology
 alternative, 12, 13, 38, 124-125, 127, 129, 131, 132, 138, 139, 141, 142, 144, 145, 147, 153, 159, 168, 169, 170, 175, 176, 179, 203
 augmenting technical change, 179, 193
 base, 1, 17, 18, 24, 32, 34, 37, 38, 55
 economics, 2, 3, 5
 energy, 2, 7, 12, 112, 114-116, 121, 125-134, 137-139, 143-145, 146, 157-164, 166, 170, 174, 175, 180, 182, 191, 192
 essential, 13, 32, 142, 177, 178, 180, 182, 183, 192
 exploration, 1, 17, 19, 43
 extraction, 2, 5, 6, 17-19, 58, 144
 input, 13, 32, 112, 161, 180, 192, 193
 price, 1, 5, 17, 19, 34, 319
 shortage, 5, 13, 15, 124, 141, 157
 technical, 179, 193
revenue models, target, 70
Reynolds, Douglas B., 17, 37, 39, 41, 43,45, 46, 56, 57, 58, 68, 69, 95, 96, 98, 104, 110, 111, 131, 132, 140, 141, 147, 148, 157, 158, 159, 161, 171, 176, 177, 191, 195, 197
Ricardian scarcity, 3, 4, 19. *See* Malthusian scarcity and scarcity
Richards, F.J., 31, 39
risk
 aversion, 2, 9, 10, 69, 71, 75, 79, 80, 87-92, 96, 104, 106, 108, 109, 110. *See also* OPEC
 subjective, 87
 risk-loving behavior, 10, 71, 76, 77, 79
 utility, concave, 76, 80
 utility, convex, 76, 80
Roberts, F., 58, 68
Robinson Crusoe Model, 22-29, 32. *See also* mineral economy model
 consumption, 32
 corn economy model, 22
 hardtack economy model, 2,5, 22-30, 32
 information capital, 27-33
 shipwrecks, 23, 24, 25
Rocks, L, 137, 140, 148, 158
Rosenburg, N., 157, 158
roulette, 20, 21, 24, 26, 60, 61, 62. *See also* Monte Carlo simulation and random search
Russia, 57, 150, 197, 198, 202, 205, 209, 211, 212, 213, 118, 219. *See also* productivity, scarcity and Soviet Union
Ryan, J.M., 43, 56, 97, 110
Ryzhkov, Nikolai, 219

S

Salahor, G., 78, 93
Salameh, Mamdouh G., 110
Samiei, H., 98, 110
Saudi Arabia, 9, 25, 73, 74, 82, 88, 89, 90, 91, 96, 104, 106, 107, 109, 126, 205. *See also* OPEC and price leader
Savage, L.J., 69, 71, 75, 76, 93
scarcity, 1-5, 13-15, 126, 136-137, 142, 158, 171-172, 177-179, 195. *See also* Malthusian scarcity and Ricardian scarcity
 alternative theories of, 1, 2, 4, 5, 6, 9
 and exploration game, 57-58, 62, 67, 68
 and Hubbert curve, 17-19, 22-23, 35, 39
 US and Russia, 197, 199, 203, 204, 206, 210-212, 214, 218, 219
 US supply, 42-43, 56
 world supply, 92, 109-110
scenario B, 107. *See also* Hubbert curve
Scott, Anthony, 18, 39
sequence, small-large-small. *See* small-large-small sequence
shale oil, 12, 117, 131, 132, 136-40, 145, 146, 148-50, 157, 171
Sherman, George, 8, 9, 15, 20, 38, 59, 60, 61, 68
shortage, energy, 2, 5, 7, 10, 12, 13, 15, 96, 109, 124, 137, 141, 142, 144, 146, 152, 153, 157, 158, 205, 208, 217, 218
 of oil, 10, 12, 96, 109, 124, 137, 142, 144, 146, 152, 153, 157, 208, 217, 218
 of resources, 2, 5, 7, 13, 15, 141, 157, 158, 205
Simon, Julian L., 14, 15, 18, 39, 111, 158, 159, 176
Six Day War, 105
Skinner, B.J., 194
Slade, M., 58, 68, 98, 110
small-large-small sequence, 30, 43, 57, 65, 67, 68. *See also* Hubbert curve and supply
Smil, V., 181, 195
Smith, James L., 25, 39
Smith, V. Kerry, 15, 195
Socolow, R.H., 195

solar. *See also* energy and supply
energy, 12, 116, 117, 124, 127, 134, 154, 163, 164
panels, 173
power, 2, 116, 134, 145, 163, 173, 181
system, 144
vehicles, 134, 154, 163
Solow, R.M., 140, 177, 178, 183, 195
Sonntag, R.E., 193, 195
Soviet Union. *See also* capital, Communism, command economy, Council for Mutual Economic Assistance (CMEA), Eastern Europe, *glasnost,* Gorbachev, investment, oil, *perestroika*, regulation, Russia, reserves, statistical analysis, and technology
black market, 138, 211, 212, 216, 217, 218
cause and effect scenario, 198, 202, 208, 218
collapse, 2, 14, 15, 197-220
conservation, 204, 205
consumption of energy, 197, 198, 199, 203, 204, 206-08, 215, 218
defense industry, 14, 15, 201, 202
discontent, 200, 201, 207, 216
economy 14, 15, 197, 198, 199
exploration for oil, 210
growth, 199, 200, 202, 203, 204, 219
industrialization, 199, 200, 202
investment, 201, 205, 209
military buildup and expenditure, 201, 202
oil production, 202-214
price shock, 157, 198, 204, 205, 213-214, 218
stagnation, 199, 201, 202, 205
technology, 199, 200, 202, 203
transition, 197-220
Stackelburg theory, 90
stagnation, economic, 111. *See also* Soviet Union
Starr, Gary, 130
Startzman, Richard A., 43, 46, 55
state grade. *See* grade, energy
statistical analysis
alternative energy sources, 38
biomass energy, 181
energy, 126, 176, 214
Hubbert model, 41, 42, 55, 95-99, 105-110
oil consumption, 203, 215
oil price, 10, 203, 207. *See also* price and price shock
oil production, 52, 72, 96, 97, 100, 218
oil supply, 10, 38, 42, 50, 56, 95, 96, 98-102, 104, 105-08, 110, 146. *See also* reserves, scarcity, Soviet Union and United States
steel, 129, 134, 164, 179, 182, 191, 192
steam. *See also* grades, fuel, energy
Carnot cycle, 193
cycle generator, 172
engines, 166
locomotives, 151, 166, 168, 172
pressure, 193
thermal cycle, 168
steamships, 166
Stiglitz, J.E., 13, 15, 178, 195
stock. *See* fixed stock paradigm
structural resources, 180-82, 186, 191-93
substitutability, 177, 178, 194
substitutes, 2, 4, 12, 13, 18, 35, 37, 54, 109, 126, 127, 138, 170, 173, 175. *See also* limits to substitution, rate of technical substitution
capital and labor, 177-95
capital-energy ratio, 187-188, 194-195
capital-labor ratio, 184-86, 189-92
capital-land ratio, 184-86, 189-92
capital-pollution ratio, 190-91
energy-capital ratio, 187-188, 194-195
energy-labor ratio, 188-189
for energy, 12, 190
labor-capital ratio, 184-86, 189-92
labor-energy ratio, 188-89
labor-land ratio, 189-90
land-capital ratio, 189-90
land-labor ratio, 189-90
for oil, 100, 145-52, 157
transformation capital-energy ratio, 188
supply, 7, 9, 20, 23, 24, 33, 35, 36, 39, 48, 53-56, 92, 95-110, 122, 123, 127, 141, 146, 153, 178, 182, 208, 212, 219. *See also* price, price shock and scarcity
elasticity of, 7, 45, 53, 99
limit, 7, 45, 98, 99, 101, 105, 108
oil, 10, 38, 41-44, 55, 56, 68, 95-100, 105-108, 110, 208, 219
trend, 35
Suranson, V.E., 140
sustainability. *See* technology
sustainable development, 158, 177-79, 195
sustainable growth, 187
sustainable world, 158, 176
synthetic fuels, 147-49, 157

T

tar sands, 146, 148, 150
target revenue models, 70
technical rate of substitution, 183, 186, 187
technology, 1, 2, 4, 5, 7, 8, 11, 13-15, 17-19, 22, 23, 25, 33-38, 43, 45-47, 49, 55, 58, 59, 68, 97, 98, 100, 103, 109-112, 120-124, 127, 128, 132-142, 146, 147, 150, 152, 153, 156-164, 166, 167, 170, 171, 175, 178, 179, 181, 182, 187, 190, 192-194, 199-203, 205, 208, 209, 211-217. *See also* energy, epochal innovation, Soviet Union, supply, and sustainability
  alternative energy sources, 11-14, 141-47, 150, 152-54, 156-58, 160-67, 170, 171, 174-76
  auto, 156
  coal, 123, 124
  computer, 157
  diffusion, 43, 58
  electric, 154, 153
  energy, 11, 12, 144, 156, 157
  energy conservation, 145
  energy consumption, 145
  energy conversion, 145
  exploration, 144
  extraction, 144
  gas-to-liquid (GTL), 147, 148, 170, 171
  information effect, 1, 3-7, 9, 10, 17, 19-33, 36, 37, 43, 47, 50, 51, 54, 57-59, 61-64, 66-68, 78. *See also* Hubbert curve, resources, and supply
  innovation, 5, 14-18, 111, 128, 159, 160, 164, 174, 179
  oil shale, 12, 136
  overcoming scarcity, 1, 2, 4, 5, 7, 13, 14, 19, 22, 23, 25, 37, 38, 178
  research, 144
  Soviet Union, 14-15, 199, 200, 202-203, 205, 208, 209, 211-15, 217
  sustainability, 178-79, 181-82, 187, 190-92. *See also* weak sustainability
Teece, D.J., 70, 93
ten-to-one ratio. *See* Hubbert curve, reserves and supply
Tengiz oil field, Kazakstan, 213
Texas Railroad Commission (TRC), 48, 49, 53, 166
Tilton, J.E., 142, 158, 177, 195
tit-for-tat, 74, 89, 91. *See also* game theory
transformation capital-energy ratio, 188
TRC. *See* Texas Railroad Commission

U

UAE (United Arab Emirates), 104
Uhler, R.C., 4, 24, 29, 39, 98, 110
ultimately recoverable reserve (URR). *See* oil, reserve, resource, supply
Union of Soviet Socialist Republics (USSR). *See* Soviet Union
United States. *See also* productivity and scarcity
  oil supply, 42-56, 59, 62, 95-99, 103-107, 146, 204, 210-12, 214, 218. *See also* reserves and statistical analysis
  ultimately recoverable reserves. *See* reserves (URR)
United States Department of Energy (DOE). *See* Department of Energy
United States Energy Information Administration (EIA). *See* Energy Information Administration (EIA).
United States Foreign Broadcasting Information Services (FBIS). *See* Foreign Broadcasting Information Services (FBIS)
uranium, 117, 128. *See also* nuclear energy
URR. *See* reserves
U-shaped cost model, 58, 186
USSR. *See* Soviet Union
utility theory, 32, 69, 71, 75-86, 88, 93. *See also* gamble, game theory, OPEC, and risk
  of gains, 84, 88

V

VanWylen, G.J., 195
Venezuela, 150, 172
volume grade. *See* grade, energy
Von Neumann-Morgenstern utility function, 69, 93

W

Watkins, C.J., 34, 39, 49, 56
weak sustainability, 177. *See also* technology
weight grade. *See* grade, energy
Weir, Fred, 199, 202, 219
wellhead. *See* oilwells
Wellman, P., 140
Williams, Robert H., 176

Wiorkowski, John J., 31, 39, 97, 110
winning, 21, 60, 63, 64. *See also* game theory model and Monte Carlo simulation
wood, 11, 114, 115, 117, 118, 121-24, 132-34, 138, 155, 162, 164, 165, 169, 180, 181, 191, 192, 193. *See also* coal, energy, England, fuel, and grades

X

Xiong, W., 93

Y

yield per effort (YPE), 6, 7, 44
Young, W.J., 158
YPE. *See* yield per effort
Yucel, M., 70, 72, 92

SYMPOSIUM SERIES

1. Jïrgen Moltmann *et al.*, **Religion and Political Society**
2. James Grace, **God, Sex, and the Social Project: Glassboro Papers on Religion and Human Sexuality**
3. M. Darroll Bryant and Herbert W. Richardson (eds.), **A Time for Consideration: A Scholarly Appraisal of the Unification Church**
4. Donald G. Jones, **Private and Public Ethics: Tensions Between Conscience and Institutional Responsibility**
5. Herbert W. Richardson, **New Religions and Mental Health: Understanding the Issues**
6. Sheila Greeve Davaney (ed.), **Feminism and Process Thought: The Harvard Divinity School / Claremont Center for Process Studies Symposium Papers**
7. A.T.D./Fourth World, **Children of Our Time: Words and Lives of Fourth World Children**
8. Jenny Yates Hammett, **Woman's Transformations: A Psychological Theology**
9. S. Daniel Breslauer, **A New Jewish Ethics**
10. Darrell J. Fasching (ed.), **Jewish People in Christian Preaching**
11. Henry Vander Goot, **Interpreting the Bible in Theology and the Church**
12. Everett Ferguson, **Demonology of the Early Christian World**
13. Marcia Sachs Littell, **Holocaust Education: A Resource Book for Teachers and Professional Leaders**
14. Char Miller, **Missions and Missionaries in the Pacific**
15. John S. Peale, **Biblical History as the Quest for Maturity**
16. Joseph Buijs, **Christian Marriage Today: Growth or Breakdown**
17. Michael Oppenheim, **What Does Revelation Mean for the Modern Jew?**
18. Carl F.H. Henry, **Conversations with Carl Henry: Christianity for Today**
19. John Biermans, **The Odyssey of New Religious Movements: Persecution, Struggle, Legitimation. A Case Study of the Unification Church**
20. Eugene Kaellis and Rhoda Kaellis (eds.), **Toward a Jewish America**
21. Andrew Wilson (ed.), **How Can the Religions of the World be Unified?: Interdisciplinary Essays in Honor of David S.C. Kim**
22. Marcia Sachs Littell *et al.*, (eds.), **The Holocaust Forty Years After**
23. Ian H. Angus, **George Grant's Platonic Rejoinder to Heidegger: Contemporary Political Philosophy and the Question of Technology**
24. George E. Clarkson, **Grounds for Belief in Life After Death**
25. Herbert W. Richardson, **On the Problem of Surrogate Parenthood: Analyzing the Baby M Case**

26. Leslie Muray, **An Introduction to the Process-Understanding of Science, Society, and the Self: A Philosophy for Modern Man**
27. Dan Cohn-Sherbok (ed.), **The Salman Rushdie Controversy**
28. James Y. Holloway (ed.), **Barth, Barmen and the Confessing Church: *Katallegate***
29. Ronald F. Duska (ed.), ***Rerum Novarum*–A Symposium Celebrating 100 Years of Catholic Social Thought**
30. Franklin H. Littell, Alan L. Berger, and Hubert G. Locke (eds.), **What Have We Learned?: Telling the Story and Teaching the Lessons of the Holocaust: Papers of the 20th Annual Scholars' Conference**
31. Alan L. Berger (ed.), **Bearing Witness to the Holocaust, 1939-1989**
32. G. Jan Colijn and Marcia S. Littell (eds.), **The Netherlands and Nazi Genocide: Papers of the 21st Annual Scholars' Conference**
33. Marion Mushkat, **Philo-Semitic and Anti-Jewish Attitudes in Post-Holocaust Poland**
34. Katharine B. Free (ed.), **The Formulation of Christianity by Conflict Through the Ages**
35. W. Anthony Gengarelly, **Distinguished Dissenters and Opposition to the 1919-1920 Red Scare**
36. John J. Carey (ed.), **The Sexuality Debate in North American Churches 1988-1995: Controversies, Unresolved Issues, Future Prospects**
37. James P. Hurd (ed.), **Investigating the Biological Foundations of Human Morality**
38. Allen E. Hye, **The Moral Dilemma of the Scientist in Modern Drama: The Inmost Force**
39. Will Morrisey, **A Political Approach to Pacifism** (2 Volumes - 39A-39B)
40. LaRae Larkin, **The Legitimacy in International Law of the Detention and Internment of Aliens and Minorities in the Interest of National Security**
41. Brian L. Fife, **School Desegregation in the Twenty-First Century: The Focus Must Change**
42. Paul Leslie (ed.), **The Gulf War as Popular Entertainment: An Analysis of the Military-Industrial Media Complex**
43. Jutta T. Bendremer, **Women Surviving the Holocaust: In Spite of the Horror**
44. George Anastaplo, **Campus Hate-Speech Codes, Natural Right, and Twentieth Century Atrocities**
45. David Brez Carlisle, **Human Sex Change and Sex Reversal: Transvestism and Transsexualism**
46. Jack Parsons, **Human Population Competition: A study of the pursuit of power through numbers** (2 Volumes - 46A-46B)
47. Adam Neuman-Nowicki, **Struggle for Life During the Nazi Occupation of Poland**, translated and edited by Sharon Stambovsky Strosberg

48. Christopher G. Hudson, **An Interdependency Model of Homelessness: The Dynamics of Social Disintegration**
49. Michèle Chaban, **The Life Work of Dr. Elisabeth Kübler-Ross and Its Impact on the Death Awareness Movement**
50. Englebert Ssekasozi, **A Philosophical Defense of Affirmative Action**
51. Kimmo Kääriäinen, **Religion in Russia After the Collapse of Communism: Religious Renaissance or Secular State**
52. Edward Timms and Andrea Hammel (eds.), **The German-Jewish Dilemma: From the Enlightenment to the Shoah**
53. Victoria La' Porte, **An Attempt to Understand the Muslim Reaction to the Satanic Verses**
54. Elizabeth S. Spragins, **Metaphoric Analysis of the Debate on Physician Assisted Suicide**
55. Alice Duhon-Ross (ed.), **Reaching and Teaching Children Who Are Victims of Poverty**
55. Alice Duhon-Ross (ed.), **Reaching and Teaching Children Who Are Victims of Poverty**
56. Joan C. McClennen and John Gunther (eds.), **A Professional's Guide to Understanding Gay and Lesbian Domestic Violence: Understanding Practice Interventions**
57. Seamus Dunn and Helen Dawson, **An Alphabetical Listing of Word, Name and Place in Northern Ireland and the Living Language of Conflict**
58. Alexandre Kimenyi and Otis L. Scott (eds.), **Anatomy of Genocide: State-Sponsored Mass-Killings in the Twentieth Century**
59. Margaret Phipps Brown, Samuel L. Dameron, Richard H. Moore, **The Role of Firearms in Domestic Violence: A Study of Victims, Police, and Domestic Violence Shelter Workers in West Virginia**
60. Patricia Ainsa, **Teaching Children With AIDS**
61. J. Sanford Rikoon and Theresa L. Goedeke, **Anti-Environmentalism and Citizen Opposition to the Ozark Man and the Biosphere Reserve**
62. Jules Steinberg, **Hannah Arendt on the Holocaust: A Study of the Suppression of Truth**
63. Stephen Gilliatt, **An Exploration of the Dynamics of Collaboration and Non-Resistance**
64. Lola M. Butler, Diane Elliott, John Gunther (eds.), **Strategies to Overcome Oppression and Discrimination for Marginalized Groups**
65. Douglas B. Reynolds, **Scarcity and Growth Considering Oil and Energy: An Alternative Neo-Classical View**
66. Steven L. Spiegel, Jennifer D. Kibbe, Elizabeth G. Matthews (eds.), **The Dynamics of Middle East Nuclear Proliferation**
67. Darlington I. I. Ndubuike, **The Struggles, Challenges and Triumphs of the African Immigrants in America**